T0256683

What Is a Complex System?

What Is a Complex System?

J. Ladyman and K. Wiesner

Yale UNIVERSITY PRESS

New Haven & London

Copyright © 2020 by James Ladyman and Karoline Wiesner.

All rights reserved.

This book may not be reproduced, in whole or in part, including illustrations, in any form (beyond that copying permitted by Sections 107 and 108 of the U.S. Copyright Law and except by reviewers for the public press), without written permission from the publishers.

Yale University Press books may be purchased in quantity for educational, business, or promotional use. For information, please e-mail sales.press@yale.edu (U.S. office) or sales@yaleup.co.uk (U.K. office).

Printed in the United States of America.

Library of Congress Control Number: 2020930179
ISBN 978-0-300-25110-4 (pbk: alk. paper)

A catalogue record for this book is available from the British Library.

This paper meets the requirements of ANSI/NISO Z39.48-1992 (Permanence of Paper).

10 9 8 7 6 5 4 3 2 1

Contents

Preface   **ix**

1 Introduction   **1**
   1.1 What Is a Complex System?   9
   1.2 A Brief History of Complexity Science   11
      1.2.1 Cybernetics and Systems Theory 12
      1.2.2 Dynamical Systems Theory 13
      1.2.3 Cellular Automata 15
      1.2.4 The Rise of Complexity Science 16

2 Examples of Complex Systems   **19**
   2.1 Matter and Radiation 19
   2.2 The Universe 29
   2.3 The Climate System 33
   2.4 Eusocial Insects 37
      2.4.1 Ant Colonies 38
      2.4.2 Honeybee Hives 41
   2.5 Markets and Economies 45
   2.6 The World Wide Web 54
   2.7 The Human Brain 57

3 Features of Complex Systems   **63**
   3.1 Numerosity 66
   3.2 Disorder and Diversity 68
   3.3 Feedback 70
   3.4 Non-Equilibrium 71
   3.5 Interlude: Emergence 73
   3.6 Order and Self-Organisation 76
   3.7 Nonlinearity 77
   3.8 Robustness 79

3.9   Nested Structure and Modularity   81
3.10 History and Memory   81
3.11 Adaptive Behaviour   82
3.12 Different Views of Complexity   84

4   Measuring Features of Complex Systems   87
4.1   Numerosity   88
4.2   Disorder and Diversity   89
4.3   Feedback   93
4.4   Non-Equilibrium   95
4.5   Spontaneous Order and Self-Organisation   96
4.6   Nonlinearity   99
       4.6.1   Nonlinearity as Power Laws   99
       4.6.2   Nonlinearity versus Chaos 100
       4.6.3   Nonlinearity as Correlations or Feedback 101
4.7   Robustness 101
       4.7.1   Stability Analysis 102
       4.7.2   Critical Slowing Down and Tipping Points 103
       4.7.3   Self-Organised Criticality and Scale Invariance . . . 105
       4.7.4   Robustness of Complex Networks 106
4.8   Nested Structure and Modularity 107
4.9   History and Memory 110
4.10 Computational Measures 110
       4.10.1 Thermodynamic Depth 111
       4.10.2 Statistical Complexity and True Measure
               Complexity 111
       4.10.3 Effective Complexity 114
       4.10.4 Logical Depth 115

5   What Is a Complex System?   117
5.1   Nihilism about Complex Systems 117
5.2   Pragmatism about Complex Systems 119
5.3   Realism about Complex Systems 120
       5.3.1   Generic Conceptions of Complexity 121
       5.3.2   Physical Conceptions of Complexity 122
       5.3.3   Computational Conceptions of Complexity 124
       5.3.4   Functional Conceptions of Complexity 124
5.4   The Varieties of Complex Systems 126
5.5   Implications 132

Appendix – Some Mathematical Background   **135**

   A   Probability Theory   135

   B   Shannon Information Theory 137

   C   Algorithmic Information Theory 139

   D   Complex Networks 140

Bibliography   **143**

Index   **163**

Preface

This book is the result of a decade-long collaboration that began in 2007 with the opening of the Centre for Complexity Sciences at the University of Bristol (an EPSRC-funded doctoral training centre). Teaching at this centre brought us together, and we began discussing the different phenomena and measures of complexity science and the ideas associated with it. We realised the need for a thorough analysis which would answer questions such as 'Is complexity a truly new phenomenon or merely a new label?'; 'Can the different conceptions of complexity of physicists, biologists, social scientists and others be brought into a single framework, or do they address different and unrelated phenomena?'; 'Are measures of complexity meaningful for a phenomenon this multi-faceted?'; and 'Why are information theory and network theory so prominent in complexity science?'. We strongly believed that, whatever the answers to these questions are, it would be beneficial to define precisely the terms of the debate, the phenomena they describe, and the relations between these phenomena. Lack of clarity in these respects is detrimental to science, and confused foundations are sometimes highly problematic. We came to the conclusion that the field of complexity sciences has harboured a lot of confusion, perhaps because it is rather young and includes so many different branches. Our first work on this problem led to Ladyman, Lambert, and Wiesner (2013), upon which this book builds.

This book is written for students and academics interested in complexity science and the nature of complexity and for other scientific practitioners in related areas, as well as for scientifically informed general readers. We have striven for conceptual and linguistic clarity and precision throughout and have sought to make our ideas and reasoning as simple as possible, while always being scientifically accurate. We explain both the foundations of complexity and the mathematical and computational tools currently being used by complexity scientists.

The first chapter provides an overall introduction to the subject and a brief account of its history and that of related fields. This chapter is very widely accessible and contains much that will be familiar to complexity scientists, though even they will find it worthwhile to read how we formulate the ideas and issues. We have found that sense can be made of most of what has been written by complexity scientists about complexity and that most of what is claimed is for good reason. However, sense and reason can be obscured by the words used to express them. We hope some of what we say in this book seems obvious in hindsight because it formulates clearly and exactly what many people already know. For example, the 'truisms' of complexity science are unlikely to be disputed by any expert in complexity science, though they have never been stated this clearly and explicitly before.

The second chapter reviews typical examples of complex systems and shows how diverse they are and the wide range of features that they display, as well as some of their commonalities. This chapter is also largely very widely accessible, though some parts presuppose some knowledge of physical science. Again we think that experts will find much of what we say uncontroversial, though some will dispute that all the examples we discuss are genuine examples of complex systems. Our method in the rest of the book is to answer the questions raised briefly above and more fully in the first chapter by using the examples we discuss in the second chapter as data for our conceptual analysis.

Chapter 3 and Chapter 4 are the core of our account of the foundations of complexity. In the former we compose a list of features of complex systems, and in the latter we relate this to the measures used by scientists in the field. Some of the material in Chapter 4 requires knowledge of mathematics as we explain many mathematical and computational tools and their role in studying complexity and refer to the relevant scientific literature. The final chapter argues for our own view of the consequences of the analysis of the preceding chapters for the notion of a complex system and the phenomena of complexity, as well as for the status of complexity science as a discipline. Parts of this chapter rely on the discussions of the previous chapters, but we summarise their conclusions so that the reader who did not follow all the details can still follow our reasoning. We do not engage much directly with the extensive philosophical literature on emergence and reduction as that would require a book unto itself. Philosophers have defined many versions of both ideas, and we have adopted the most simple taxonomy as explained in the text.

We have many people to thank for their help and support over the years, for their insightful comments on drafts of this book, and for discussions of the subject. In particular, we would like to thank the students of the Bristol Centre for Complexity Sciences for many discussions on the subject and for their feedback on early drafts of this book. We would like to thank our colleagues at the University of Bristol and elsewhere. Colleagues who have been particularly helpful with comments on the manuscript are Colm Connaughton, Doyne Farmer, Mauro Fazio, Alasdair Houston, Jenann Ismael, Christopher Jones, Gordon McCabe, Melanie Mitchell, Samir Okasha, Stuart Presnell, Don Ross, Anne-Lene Sax, Danny Schmidt, Karim Thébault, Lucy Viegas, Thorsten Wagener, Jim Weatherall, and Lena Zuchowski as well as a number of anonymous referees. We thank Elisa Bozzarelli for designing the Figure in Chapter 5.

Chapter 1

Introduction

Complexity science is relatively new but already indispensable. It is important to understand complex systems because they are everywhere. Your brain is a complex system and so is your immune system and every cell in your body. All living systems and all intelligent systems are complex systems. The climate of the Earth is a complex system, and even the universe itself exhibits some of the features of complex systems. Many of the most important problems in engineering, medicine and public policy are now addressed with the ideas and methods of complexity science – for example, questions about how epidemics develop and spread. Thousands of years of mathematical and scientific study have given us the technology to create new complex systems that rival those of the biosphere, such as cities, financial economies and the Internet of Things. Business leaders have started to think in terms of complexity science, using terms such as 'robustness', 'redundancy' and 'modularity' (Sargut and McGrath 2011; Sullivan 2011). State economic institutions such as the Bank of England (Haldane 2009) have also begun to use such terminology. This book is about how scientists think about complex systems and about what makes these systems special.

However, there is confusion in some of the discussions in the professional and scientific literature, and clarity is needed to facilitate the application of complexity science to problems in science and society. There is no agreement about the definition of 'complexity' or 'complex system', nor even about whether a definition is possible or needed. The conceptual foundations of complexity science are disputed, and there are many and diverging views among scientists about what complexity and complex systems are. Even the status of complexity as a discipline can be questioned given that it potentially covers almost everything.

Most sciences admit of informative definitions that are easy to state. For example, biology is the study of living systems, chemistry is the study of molecular structure and its transformations, economics is the study of the allocation of scarce resources that have different possible uses, and physics is the study of the most basic behaviour of matter and radiation. Complexity science is the study of complex systems, and, while it may be difficult exactly to define 'life', 'matter' and the other things just mentioned, to say what complex systems are is even harder. There is no agreement about what complexity is, whether it can be measured, and, if so, how, and there is no agreement about whether complex systems all have some common set of properties.

There are examples that everyone agrees are complex systems, but there are also many disputed cases. For example, some people regard a purely physical system like the solar system as a complex system (Simon 1976), while others think that complex systems must display adaptive behaviour (Holland 1992; Mitchell 2011), so only systems that have functions and goals can be complex.[1] The rest of this section states clearly what can be said about complexity science that is not contentious, beginning with the limitations of the rest of science that make it necessary.

Knowledge of physics and chemistry has enabled us to control many aspects of the world. The fundamental laws of mechanics and electromagnetism have a beautiful simplicity and incredible predictive accuracy. The atomic theory of matter, according to which all the material things we see around us are composed of elements like carbon and oxygen, can be used to understand the physical components of every chemical substance. However, many phenomena are very messy, and the behaviour of many systems, even relatively simple ones, is very hard to describe in detail. For example, the flow of turbulent water and the formation of a snow crystal are incredibly intricate phenomena involving a huge number of variables (a single snow crystal contains around 10^{18} molecules). Although fantastic progress has been made in computation and simulation, measuring and calculating the state of every molecule in a real snow storm is not remotely feasible.

Furthermore, the physics and chemistry of atoms and molecules cannot be used to predict individual people's actions, where the stock market will be tomorrow, or what the weather will be next week, because they cannot be directly applied to such problems at all. People, markets,,the atmosphere, and their properties are described by psychology, economics and climatology respectively. Even within physics there are many levels of description of entities and processes at very different length and time scales, from the

---

[1]The term used in much of the literature is 'complex adaptive behaviour', but we drop the word 'complex' . A similar point was made by Murray Gell-Mann (1994, p. 27).

protons and electrons in the standard model of particles, to stars and galaxies in astrophysics. There is a lot of science that links the phenomena at different scales. For example, quantum chemistry links chemical reactions to the electromagnetic interactions between subatomic particles, and the kinetic theory of gases links the pressure and temperature of gases to the collisions and motions of their molecules. However, it is impossible to describe the solar system just using fundamental physics.

In general, collections of things can have different kinds of properties to their parts. For example, properties like pressure do not pertain to individual molecules but to gases. In a macroscopic sample of a gas, there are billions upon billions of molecules and many collisions and motions. If the gas is in a sealed container, then all these processes automatically make the gas approximately obey three laws. One of them is Boyle's law, stating that the pressure is inversely proportional to the volume at a fixed temperature. These 'ideal' gas laws relate the properties of pressure, volume and temperature independently of the kind of gas and regardless of the exact and incredibly complicated behaviour of the particles, all of which are extremely fast and short-lived compared to the time scale of the behaviour of the whole gas. (They are called 'ideal' because real gases do not obey them exactly.) Sometimes systems obey laws that are general and allow us to neglect almost all the details, and in this way simplicity can come from something very complicated.

There is no need to believe that some mysterious new ingredient has to be added to molecules to make gases. Gases and their properties are the result of the relations and interactions among the parts of the gas. If the sum of the parts is taken to be just the collection of the parts as if they were in isolation from each other, then the whole is more than the sum of the parts. However, the interactions of the parts are all it takes to make the whole exist. One of the most fundamental ideas in complexity science is that the interactions of large numbers of entities may give rise to qualitatively new kinds of behaviour different from that displayed by small numbers of them, as Philip Anderson says in his hugely influential paper, 'more is different' (1972).

When whole systems spontaneously display behaviour that their parts do not, this is called 'emergence'. Even relatively simple physical systems, such as isolated samples of gases, liquids and solids, display emergent phenomena in the minimal sense that they have properties that none of their individual molecules have singly or in small numbers. However, there are many different kinds of emergence that are much more intricate – for example, when systems undergo 'phase transitions', such as turning from liquid to solid or from insulator to superconductor. Phase transitions and associated 'critical

3

phenomena' are examples of spontaneous self-organisation, in which physical systems are driven from the outside and there is emergent order to their behaviour. Systems can be driven by heat, for example, and also by a flow of matter. The famous Belousov-Zhabotinsky reaction produces patterns of different coloured chemicals that oscillate as long as more reagents are added. Such examples show that there are many rich forms of emergent behaviour in nonliving systems and that *nonliving systems can generate order*.[2] (These examples and those of some living systems mentioned below are explained in Chapter 2.)

Biological systems display many further examples of emergence, including metabolism and the coding for proteins in DNA, the representation of the state of the environment by perceptual systems, and adaptive behaviour like foraging and the rearing of offspring. Emergence in collectives of organisms includes the social behaviour found, for example, in beehives and ant colonies, which are in some ways like a single meta-organism, and elephant herds and primate groups, whose societies can be very sophisticated. There are cases of collective motion being directed by a privileged individual, such as a herd of horses following the leading mare. However, a flock of birds moves as a whole without a special individual leading it. Similarly, when social insects make decisions, such as bees collectively flying off to a new nest, they do so without one individual playing any special role in the group. Instead, their collective behaviour arises just as a result of their interactions and the feedback between their responses to each other's behaviour. A central idea in complexity science is that complex systems are spontaneous products of their parts and the interactions among them. Individual ants and small numbers of them just wander around aimlessly, but in large numbers they build bridges, maintain their nests and even grow fungi in them. This is another of the lessons of complexity science. *Coordinated behaviour does not require an overall controller*.

There is sometimes an underlying simplicity to the production of coordination and order that can be put in mathematical terms. It is surprising that the collective motion of a flock of birds, a shoal of fish, or a swarm of insects can be produced by a collection of robots programmed to obey just a couple of simple rules (Hamann 2018). Each individual must stay close to a handful of neighbours and must not bump into another individual. It regularly checks how close it is to others as it moves and adjusts its trajectory accordingly. As a result, a group moving together spontaneously forms. The adaptive behaviour of the collective arises from the repeated interactions, each of which on its own is relatively simple. This is another conclusion of complexity

---

[2]For a discussion of emergence in physics see Butterfield (2011a,b).

science: *complexity can come from simplicity.*[3]

The ideal gas laws mentioned above as a very simple example of emergent behaviour are of limited application because they do not apply to real gases under many circumstances (for example, at very low temperatures or very high density or if a gas is compressed very quickly, heated very rapidly, or suddenly allowed to expand). Similarly, all sciences involve ways of approximating, idealising and neglecting details. For example, the law of the pendulum, which says that the time period of oscillation depends on the length of the string but not on the mass of the bob, applies only when the line connecting the bob to the pivot can be treated as being massless because it is so small compared to the mass of the bob. Similarly, Newton was able to work out the inverse square law of gravitation only because there is negligible friction to affect the motion of the planets, and their attraction for each other is negligible compared to the attraction of the sun. Even in such ideal circumstances, the equations describing how more than two bodies behave in general cannot be solved exactly and numerical methods must be used, or a restricted class of systems must be studied (Goldstein 1950).

Knowing how to model any complex system requires knowing what idealisations and approximations to make. Complexity science involves distinctive kinds of approximation and idealisation. For example, the Schelling model of segregation treats a population and its residences as a lattice of squares, each of which can be populated or not by one of two types of individuals (Schelling 1969). The system evolves according to the rule that individuals move on a given turn if and only if they are surrounded by fewer individuals of the same type than some specified number. The stable states of such systems are highly segregated, and in them most individuals are surrounded by others of the same type. These models show that segregation can arise even when individuals have a relatively mild preference for being near others they perceive to be in some way similar to themselves. This model can be applied not just to residence, but also to the formation of social networks (Henry et al. 2011).

Sometimes multiple approximations can be made, and different models of the same system often suit different purposes. For example, the nucleus of an atom can be modelled as a liquid drop, for the purpose of studying its overall dynamics, or with its component particles occupying shells analogous to those used to describe the atomic orbitals of electrons, for studying how it interacts with radiation. Similarly, there are very diverse models in complexity science. For example, economic agents can be modelled as computational agents whose states are updated according to rules describing flows of infor-

---

[3]Strevens (2016) discusses the relationship between complexity and simplicity.

mation or as nodes in a network that are connected if two agents trade with each other. *Complex systems are often modelled as networks or information-processing systems.*

Complex networks can represent vastly different types of systems and the connections in a network may represent interactions of various kinds. For example, both the human body and a city can be modelled as a network with links representing the flow of energy, food and waste between many sites. However, networks do not represent only the flow of matter or energy, but also of information, causal influence, communication, services, or activation (among other things). In network models, the exact nature of the interactions may even be ignored when the properties of the system that are directly studied are the connections and interactions among the parts considered abstractly (Easley and Kleinberg 2010). In the biological and behavioural sciences models can be highly abstract – for example, graphs that only show ancestry relations – and highly idealised – for example, models of markets that treat agents as having perfect information.

While the interactions between the components in a network have some particular nature and are governed by the corresponding laws, often we can ignore the details about them, because the complex behaviour depends only on more abstract features of the interactions, such as how often they happen and between which parts. For example, in an economy, agents interact either face to face, or by post, or electronically, but how they interact is irrelevant beyond the implications for the timing and reliability of the exchange of information and resources. Similarly, each bird in a flock is an individual organism with a heart, a skin, eyes; it has an age, a certain size, and the need for food for survival and for procreation and many other things. But when scientists are studying collective motion, all that needs to be modelled is that the individuals in the group have a way of telling how close they are to each other. It is not important whether they do so by sight, like birds, or echolocation, like bats. The effect is the same, as long as they get the information somehow. Bees communicate by dancing when choosing where to make a new nest, but that is not important to the model of how the decision making occurs. Amazingly, the way your brain makes simple decisions is very similar, with neurons being analogous to bees. Such similarity is often captured by a common mathematical description of the different systems in question (more of this in Chapter 2 and Chapter 4). This is another important lesson of complexity science: *There are various kinds of universality and forms of universal behaviour in complex systems.*

Some complex systems involve billions upon billions of interactions between vast numbers of individuals. The complexity that can emerge is astonishing. Even the dynamics of the interactions of a thousand birds in a flock following two simple rules are beyond what a human being can calculate. Successful scientific modelling of the structure that can arise from repeated interaction requires computers. Without very powerful computers, it is impossible, for example, to collate all the data to map the flow of gas, electricity, water, people, and information in a city. Only for a few decades have we had the necessary computational power to analyse complex behaviour, simulate complex systems, and test hypotheses about how simple interaction rules and feedback produce complex behaviour. Even with vast computational power many complex systems are so complicated that making precise predictions about exactly what a particular system will do is practically impossible. Hence, predictions of real world complex systems are always of a statistical nature. In general, *complexity science is computational and probabilistic*.

Complexity science is often contrasted with reductive science, where the latter is based on breaking wholes into parts. This is misleading, because, as the rest of this book shows, complexity science always involves describing a system by describing the interactions and relations among its parts. The parts of complex systems interact by various mechanisms studied by individual scientific disciplines. Furthermore, the remarkable properties of complex systems arise because of the effects of the laws that govern the parts and their interactions. However, when there are many parts and they interact a lot, studying them requires other methods as well as those of the more fundamental science or sciences that describe the parts and involves new concepts and theories to describe the novel properties that the parts on their own do not display. In most complex systems, the interactions between the parts are of more than one kind. For example, there are both chemical and electrical interactions in the brain and both electromagnetic and gravitational interactions in galaxies. Hence, for these reasons, in complexity science often no single theory encompasses the system of interest.

Clearly complexity science would not be possible without the rest of science, and it involves combining theories from different domains and synthesising tools from various sciences. Complexity science does not involve revisions to fundamental laws, but it does involve the discovery of completely new implications of these laws for the behaviour of aggregates of systems that obey them. This is one reason why *complexity science involves multiple disciplines*. Scientific theories that have been studied and applied autonomously are integrated in a single context. Complexity science is therefore essentially interdisciplinary in both method and subject matter. It uses established scientific theories from whatever domain is relevant to the system at hand and then uses whatever resources are needed to combine them. Particular sciences provide different aspects of the explanation of the overall behaviour of the system. The relevant theories and the relations between them provide the basis for a new (complexity) theory of the system and new ways of explaining and predicting its features.

Complexity science combines the science specific to the kind of system being studied with mathematical theories, models and techniques from computer science, dynamical systems theory, information theory, network analysis and statistical physics. Much has been learned in this way about complex systems in neuroscience, cell biology, economics, astrophysics and many other sciences, and the techniques of complexity science are now essential for much of engineering, medicine and technology.

Understanding the nature of complex systems is made more difficult by the fact that complexity science studies both systems that produce complex structures and those structures themselves. Nature is full of beautiful patterns and symmetries, such as those of honeycombs, shells and spiderwebs, which are made by living systems. Intricate structures are also found in the nonliving world – for example, in the rings of Saturn or geometrical rock formations on Earth. *There is a difference between the order that complex systems produce and the order of the complex systems themselves.*

The most astonishing example of novel properties arising in a biological system is the human brain. Our mental life and consciousness somehow emerge from the electrical and biochemical interactions among neurons. Human beings and culture are the most complex systems of which we know, and there are layers upon layers of complexity within them: for example, the many individual actions that give rise to the single event of an election or a stock market crash; the intricate feedback between humans and the climate and the environment; and the incredible complexity of a city where millions of people live and interact from moment to moment. There are many kinds of interactions, such as business transactions, bus journeys, crimes, school classes, car crashes, and chats between neighbours. Yet simple predictable social behaviour does sometimes arise. For example, many diverse properties of cities from patent production and personal income to pedestrians' walking speed are approximated by functions of population size (Bettencourt et al. 2007).

The next section introduces the main question of this book and how to answer it. First, we repeat the core claims above.

The Truisms of Complexity Science

Truisms state the obvious. The following statements will not be obvious to everyone, but they will be to those working in complexity science. However, these truisms have not yet been stated clearly and explicitly. They are the starting point for the analysis of this book because they state the basic facts about the subject while being compatible with the very wide range of views about the nature of complexity science and complex systems found in the literature.

1. More is different.

2. Nonliving systems can generate order.

3. Complexity can come from simplicity.

4. Coordinated behaviour does not require an overall controller.

5. Complex systems are often modelled as networks or information processing systems.

6. There are various kinds of invariance and forms of universal behaviour in complex systems

7. Complexity science is computational and probabilistic.

8. Complexity science involves multiple disciplines.

9. There is a difference between the order that complex systems produce and the order of the complex systems themselves.

The truisms are all independent of each other. Number 6 is an important discovery of complexity science. Note also that numbers 5, 7 and 8 are not about complex systems themselves but the science that studies them. Numbers 6 and 9 are the least obvious and most in need of the articulation and argument given for them in Chapters 3 and 4.

1.1 What Is a Complex System?

Despite the lack of consensus about how to define complex systems and complexity, there is a core set of complex systems that are widely discussed throughout the literature. Chapter 2 presents some of these canonical examples of complex systems and highlights some of their distinctive and interesting characteristics. Then Chapter 3 discusses the concepts which are ubiquitous in the scientific literature about complexity and complex systems. Ten features associated with complex systems are identified. A distinction is made between the first four, which are conditions for complexity to arise, and the rest, which are the results of these conditions and indicative of various kinds of complexity. The examples of Chapter 2 are considered in an analysis of which features are necessary and sufficient for which kinds of complexity and complex system. The features are as follows:

1. Numerosity: complex systems involve many interactions among many components.

2. Disorder and diversity: the interactions in a complex system are not coordinated or controlled centrally, and the components may differ.

3. Feedback: the interactions in complex systems are iterated so that there is feedback from previous interactions on a time scale relevant to the system's emergent dynamics.

4. Non-equilibrium: complex systems are open to the environment and are often driven by something external.

5. Spontaneous order and self-organisation: complex systems exhibit structure and order that arises out of the interactions among their parts.

6. Nonlinearity: complex systems exhibit nonlinear dependence on parameters or external drivers.

7. Robustness: the structure and function of complex systems is stable under relevant perturbations.

8. Nested structure and modularity: there may be multiple scales of structure, clustering and specialisation of function in complex systems.

9. History and memory: complex systems often require a very long history to exist and often store information about history.

10. Adaptive behaviour: complex systems are often able to modify their behaviour depending on the state of the environment and the predictions they make about it.

Some people argue that no scientific concept is useful unless it can be measured. Many putative 'measures of complexity' have been proposed in the literature, and we review some of the most prominent in Chapter 4 (the Appendix summarises some of the mathematics used in these measures). We argue that none of them measure complexity as such, but they do measure various features of complex systems. We give examples of measures of almost all of the features of complexity listed above.

Chapter 5 considers complexity science in a wider philosophical and social context, summarising what we have learned and reflecting on it. We say what we think complex systems are, argue for our view, and draw consequences from it. We argue that a system is complex if it has some or all of spontaneous order and self-organisation, nonlinear behaviour, robustness, history and memory, nested structure and modularity, and adaptive behaviour. These features arise from the combination of the properties of numerosity, disorder and diversity, feedback and non-equilibrium. We argue that there are different kinds of complex system, because some systems exhibit some but not all of the features.

We argue that our review of the scientific literature shows that the ideas of complexity and complex systems are useful in the sense of aiding successful science. We distill what it is about complex systems that makes them hard to put in the language of the traditional disciplines and what can be gained in developing a new language for them. This language allows descriptions and prediction of complex systems and their behaviour and features that would otherwise be impossible. The complex systems discussed in this book, such as beehives, brains and the climate can be remarkably resilient, but they can also be very sensitive to disruption. Understanding them is vital for our survival. The final section of this chapter briefly reviews the history of complexity science.

1.2 A Brief History of Complexity Science

The Scientific Revolution of the sixteenth and seventeenth centuries involved many crucial developments in science, including the development of calculus and Newtonian physics. However, experimental and mathematical sciences are almost as old as human civilisation. Models of the motions of the heavenly bodies and the behaviour of physical systems on Earth that were predictive and quantitative existed long before the development of modern science. Our biology and physics incorporate the work of Aristotle and Archimedes, respectively, and, while much of chemical knowledge dates from after modern chemistry developed from alchemy in the seventeenth century, many basic chemical reactions were known to the ancients and to Arab and Chinese scholars. Babylonian astronomers worked out that the Morning Star and the Evening Star are the same heavenly body (Venus), and much medical knowledge derives from ancient and medieval research. However, complexity science is very recent, and this is no accident because, as truisms 5, 7 and 8 above make clear, complexity science needs a lot of other science.

By the beginning of the twentieth century, physics and chemistry were highly sophisticated and interconnected, and the theory of probability and advanced statistical methods were under development. The origins of complexity science lie in cybernetics and systems theory, both of which began in the 1950s. Complexity science is related to dynamical systems theory, which matured in the 1970s, and to the study of cellular automata, which were invented at the end of the 1940s. By then computer science had become established as a new scientific discipline. Computation is needed for anything but the most elementary examination of what happens when a very large number of parts and their interactions form the system of interest. There is much more about all this in the rest of this book. In what follows, the main fields that led to complexity science are briefly described and their history sketched.

1.2.1 Cybernetics and Systems Theory

The idea that systems in different scientific disciplines follow similar principles, and that models for one might be useful for others, was formulated in the theory of cybernetics. Norbert Wiener and Arturo Rosenblueth coined the term 'cybernetics' in 1947 as the theory of control and communication in living and nonliving systems. Wiener, an American electrical engineer from MIT, and Rosenblueth, a Mexican physiologist, at Harvard at the time, were intrigued by the similarities between the systems they were studying. In particular, the role played by feedback in systems as different as engineered control systems and the nervous system caught their attention. Their collaboration began as regular informal meetings of a group of scientists organised by Rosenblueth at Harvard in the 1930s to discuss the possibilities of interdisciplinary collaborations. In 1943, Rosenblueth, Wiener and their colleague Julian Bigelow published a first paper on the commonalities of engineered and biological systems, introducing the idea that biological systems use feedback in adaptation (Rosenblueth et al. 1943). Also during this period, Wiener developed ideas similar to Claude Shannon's (discussed in Chapter 4) about a mathematical theory of communication. In 1947, Wiener published his seminal book, *Cybernetics: Or Control and Communication in the Animal and the Machine* (Wiener 1961). The notions of feedback and self-organisation, now central to the study of complex systems (and to much of what follows), were already present in his and Rosenblueth's work.

'Systems theory' is an overarching term for many fields, including control theory, cybernetics, systems biology and the study of adaptive systems. It goes back to Ludwig von Bertalanffy's work and his 1968 book *General System Theory: Foundations, Development, Applications* (1969). Bertalanffy, an Austrian biologist based in Vienna, wanted to generalise the observations he had made on biological systems to systems in other sciences. His early work on a mathematical model of an organism's growth over time was published in 1934. Bertalanffy's goal throughout was to find models, universal principles and laws applying to systems in general, irrespective of their particular kind. His achievement was to offer a new paradigm for conducting science, more than a universal theory applicable to all sciences (something that is arguably also true of complexity science, as discussed in what follows). The histories of cybernetics and systems theory demonstrate that the ground for complex systems science was already prepared in the mid-twentieth century by the pioneers Wiener, Rosenblueth, Bertalanffy and others.

1.2.2 Dynamical Systems Theory

The role of dynamical systems theory in the development of complexity science is various. It is no coincidence that the two fields matured in the second half of the twentieth century since they both rely heavily on computation for visualisation and exploration. The most obvious link between them is that many complex systems are dynamical systems in the most general sense, and many of them are now modeled using the tools from dynamical systems theory.

The mathematical sciences aim to predict, accurately and precisely, the observable behaviour of the world. Predictions can be accurate even if they are limited in precision. For example, astronomers use models of the solar system to predict the future positions of heavenly bodies in the night sky and the dates and times at which eclipses will occur. Their current predictions can be accurate to the nearest second. Such predictions are based on observations of the current state combined with laws to calculate the future state. There is always some error in the measurements that give the initial data, but in many circumstances this results in a proportionate amount of error in the predictions. For example, if one calculates the path of a projectile using Newton's laws of motion, errors in the magnitude and direction of the force that launches it and its mass lead to proportionate errors in the estimation of the position at which it will land. However, the equations that govern many natural systems are nonlinear in the sense that the output is not simply proportional to the input. More precisely, a linear function f is one such that:

$$f(x+y) = f(x) + f(y) \qquad \text{(superposition)},$$
$$f(cx) = cf(x), \quad \text{where } c \text{ is any constant} \qquad \text{(homogeneity)}.$$

A nonlinear function is any function that fulfills only one or none of these conditions.[4]

Consider population growth. Suppose that every pair of individuals produces four offspring. Then, in accordance with the superposition principle above, 200 individuals will produce 400 offspring. In this case, the function describing it is linear ($f(x) = 2x$), and a large population grows in just the same way as a small one. However, real population growth is not like this because, in large populations, overcrowding causes the growth rate to decline by increasing the death rate. In this case the function describing it is nonlinear. An example is the nonlinear equation of the logistic map, $f(x) = rx(1 - x)$ with parameter r, a simple model of population dynamics and now a canonical example of a chaotic system. The nonlinearity of population growth comes about because of feedback. This gives rise to sensitive dependence on initial conditions by magnifying the effect of small changes in the initial conditions as the equation describing the system is iterated over and over again.

Chaos theory is the branch of dynamical systems theory that deals with systems whose time evolution is highly sensitive to initial conditions. Chaotic systems appear to be random, because their precise behaviour in the long run is unpredictable. However, chaotic systems are deterministic in the sense that their future state is completely fixed by the laws that govern them and their present state (Strogatz 2014). The notion of sensitivity to initial conditions is famously given poetic expression by the title of a presentation by Edward Lorenz at a meeting of the American Association for the Advancement of Science, 'Predictability: Does the Flap of a Butterfly's Wings in Brazil Set Off a Tornado in Texas?' (Lorenz 1972).

In general, the longer the time evolution under consideration, the greater the uncertainty in prediction. In the case of the weather, it is in general predictable about a week in advance at most. Chaos theory began with Henri Poincaré's work in mechanics on the motion of three massive bodies in accordance with Newton's laws of motion. He observed that a simple system can be deterministic but unpredictable in practice because of sensitivity to initial conditions. Chaotic systems are described either by differential equations or by the iteration of simple mathematical formulas, meaning that the

---

[4]If x and y are real numbers, these conditions are equivalent, and if they are real vectors, then superposition implies homogeneity, but not vice versa.

result of the calculation is substituted back into the formula, and this process is repeated many times. Remarkable progress was made with paper and pencil, but only with the advent of the digital computer was it possible to perform the vast number of calculations necessary to study chaotic behaviour in depth and to generate the now-familiar images of chaotic systems such as the Lorenz attractor.

Many nonlinear and chaotic phenomena were discovered when chaos theory had matured and complexity science was still in its infancy. However, chaotic systems are not the same as complex systems. The relationship between complexity and chaos and nonlinearity is discussed in Chapters 3 and 4.

1.2.3 Cellular Automata

Cellular automata were originally conceived by Stanislaw Ulam, a Polish-American mathematician, and John von Neumann, a Hungarian-American mathematician, in the late 1940s and early 1950s (Ulam 1952; von Neumann 1966). Von Neumann's goal in particular, which he achieved, was to find computational models that could mimic biological self-reproduction. Cellular automata are arrays of cells in one, two, or more dimensions, equipped with rules for the cell states and how they can change. The simplest cellular automata are one-dimensional; they consist of one row of cells where each cell is in one of two states (for example, one of two colours). The row of cells is initiated in some configuration of states, and the configuration is changed in discrete time steps according to a set of fixed rules. Common to all rules is that the state of each cell at the next time step depends on its current state and that of its neighbours. Depending on the exact rules, different dynamics arise. When the rows of cells, each row representing one time point, are plotted underneath as a two-dimensional grid, the structures arising can be surprisingly nontrivial.

Two-dimensional cellular automata include the so-called Game of Life (Guy and Conway 1982). The state of each cell at the next time step depends on all its eight surrounding cells. Four very simple update rules lead to nontrivial dynamics that become apparent when the consecutive configurations are displayed as a video. The name derives from the lifelike structures that seem to be forming, moving, and disappearing in the video visualisation. The Game of Life is a now famous example of the emergence of structure through simple rules and multiple interactions. It can behave chaotically but can also produce relatively stable patterns with names like 'eaters' and 'gliders', which capture how they behave over time ('eaters' appear to swallow up cells, and 'gliders' appear to move across the grid). In the 1980s the study of cellular automata gained momentum due to the increasing availability of the computational power needed to simulate them.

1.2.4 The Rise of Complexity Science

One of the first meetings that discussed the connection between cellular automata, dynamical systems theory, physics, biology, and chemistry was organised by Doyne Farmer, Tommaso Toffoli, and Stephen Wolfram in 1983 at Los Alamos National Laboratory in the United States (Wolfram 1984). Many of the phenomena that are now often associated with complex systems were observed in various models of cellular automata presented at this meeting, including dissipation, phase transitions, self-organisation, and fractal structure. Participants included James Crutchfield, Peter Grassberger, Stuart Kauffman, Christopher Langton, and Norman Margolus.

Around the same time a group of Los Alamos scientists, which included George Cowan, a former White House science advisor, decided to found a new kind of PhD-granting institution focusing on interdisciplinary science. This did not come to fruition due to lack of funding, but the group did found a research institute in Santa Fe, just down the road from Los Alamos, in 1984. The first of two founding workshops had amongst the participants Phil Anderson, Charles Bennett, Felix Browder, Jack Cowan, Manfred Eigen, Marcus Feldman, Hans Frauenfelder, Murray Gell-Mann, David Pines, Ted Puck, Gian-Carlo Rota, Alwyn Scott, Jerome Singer, Frank Wilczek, and Stephen Wolfram, in addition to various scientists from Los Alamos and from other scientific institutions (Pines 2019). The title of the first workshop was 'A Response to the Challenge of Emerging Syntheses in Science: A New Kind of Research and Teaching Institution', and it involved chemists, evolutionary biologists, psychologists and anthropologists. Transcripts of this workshop were published in 2019 (Pines 2019). In one of the discussions the physicist Norman Ramsey points out that "almost always in such discussions [of interdisciplinary subjects] you omit the oldest and most fruitful ones [syntheses], such as between physics and chemistry".[5]

It is noteworthy that 'complex system' or 'complexity' were not mentioned among the topics to be discussed. The systems of interest were taken to be those that had the potential to be the subject of new syntheses between the sciences, and emergent behaviour was considered a unifying theme (in line with the emphasis placed on emergence throughout this book). Exam-

---

[5]To this David Pines replies, "Those are emerged syntheses". The question he wanted to discuss was, "Which fields in the natural and social sciences and humanities are the ones where a synthesis is or might soon be emerging?"

ples of what was discussed include the mathematics of evolutionary theory, war from an evolutionary perspective, brain mechanisms of hallucination, and modern archaeology.[6]

'Evolution, Games, and Learning: Models for Adaptation in Machines and Nature' was the title of a workshop held in 1985, again at Los Alamos National Laboratory, organised by Doyne Farmer and Norman Packard. The aim was 'attempts at synthesis rather than reduction', to address the unknowns of issues such as the principles underlying evolution and the operation of the brain using tools from computational and dynamical systems theory (Farmer and Packard 1986). The list of speakers includes Michael Conrad, John Holland, Bernardo Huberman, Stephen Kauffman, Christopher Langton, John Maynard Smith, Norman Packard, Alan Perelsen, Peter Schuster, and Stanislaw Ulam. Many of the phenomena that are now often associated with complex systems were reported on at this meeting, including evolutionary game theory and evolutionary computation, immune system and machine learning, and neural network models of learning.

These 'syntheses of ideas' all happened around the Los Alamos Lab and were driven by the people working there and their colleagues in the United States and Europe. The Santa Fe institute, which has grown in size over the years, was the first institute explicitly dedicated to the study of complex systems. It is now one of many research institutes on complex systems around the world, with most countries in Europe, the Americas, and many in South and East Asia running at least one. The international Complex Systems Society was launched in 2004 on a European level. It became intercontinental in 2006. The next chapter considers a representative sample of the kinds of system that are studied in these research institutes.

---

[6]The initial funding request to the MacArthur Foundation suggested three broad areas of inquiry: neurophysics, consciousness, and basic physics and mathematics. The funding request was unsuccessful. Nevertheless, George Cowan and his colleagues were able to raise enough money to found in 1984 a private research institute located in an old monastery in Santa Fe.

Chapter 2

Examples of Complex Systems

This chapter surveys some of the most common examples of complex systems studied by complexity science. The first two sections consider examples from physics and chemistry that show that even nonliving systems without goals or functions have properties with a rich structure and can exhibit different kinds of self-organisation and generate order. The first section is particularly important for the rest of the book, because it explains concepts that originate in physical science but which have been applied throughout complexity science (for example, the ideas of equilibrium and phase transition are widely used outside of thermal physics, where they originated). The next section is about the universe and how it and its parts, like the solar system, display various features of complex systems. The final nonliving system considered is the Earth's climate, the physics and chemistry of which both support and are generated by life. The subsequent sections discuss living systems, which display adaptive behaviour of many different kinds and different degrees of sophistication, and two complex systems of human construction – namely, the economy and the World Wide Web. The chapter ends with the human brain, which is a supreme example of a complex system. The next chapter identifies the features of complex systems displayed by these examples.

2.1 Matter and Radiation

Physics is concerned with processes involving matter and radiation at all scales, from the interactions of subatomic particles and fields to the formation of galaxies and even the universe itself. At the beginning of modern physics, in the seventeenth century, some natural philosophers, such as René Descartes and Pierre Gassendi, revived the ancient idea that the natural world is based on matter moving around in space. They became known as the 'mechanical philosophers'. Descartes thought that space is completely full of matter, but most others believed in the vacuum, especially after the invention of the air pump. The original atomists, Leucippus, Democritus and Epicurus, imagined that everything is made of tiny particles moving in a void, but they did not have a precise mathematical account of their collisions and motions. Building on the work of Kepler, Galileo, Descartes and others, Newton formulated his laws of motion and the inverse square law of gravitation, from which could be derived incredibly accurate predictions of the motions of planets, the moon, and other heavenly bodies. He could also derive the law of the pendulum, the parabolic motion of projectiles and Galileo's law of free fall. The mathematical sophistication of his work set the standard for future physics and began the era of mathematical science. The predictions of the inverse-square law of gravitation applied to the motions of the planets are accurate to one part in 10^6, even though in the seventeenth century experimental observations were only accurate to one part in 10^3. It was not until the nineteenth century that the orbit of Mercury was measured precisely enough for better physics to be empirically testable (Penrose 2004, p. 390).

For the Newtonian, the world consists both of particles and forces, such as gravity and magnetism, that push and pull these particles around. This is a departure from the mechanical philosophy whose advocates distrusted the mysterious idea of force. However, there were too many successful novel predictions, such as the bulging of the Earth at the equator and the return of Halley's comet, for such concerns to prevent its widespread adoption. It seemed that all matter might be just collections of particles and that all the phenomena we observe arise from the forces they exert on each other and their myriad motions. However, in the mid-nineteenth century, Maxwell's theory of electromagnetism incorporated fields, which came to be thought of as just as real as particles. Electromagnetic radiation and electronics are fundamental to everyday technology, such as mobile phones, microwaves and radar. Our best current physics describes the behaviour and interactions of matter and radiation in terms of quantum fields that notoriously have very strange properties and that are not at all like little bits of ordinary matter. Furthermore, the Newtonian view of space and time as fixed backgrounds to physical events has been replaced by General Relativity, which describes gravity in terms of a spacetime that is curved and dynamical.

However, classical physics is still applicable to much of what occurs in nature, even in some surprising ways. For example, colliding galaxies can be modelled as if they were gases, with stars being analogous to molecules, and the nucleus of an atom can be modelled as a liquid drop (see, for example, Rohlf 1994, Section 11.3). The statistical methods developed to describe the vast numbers of particles in macroscopic samples of matter have been applied to the statistical properties of completely different kinds of systems, such as neural networks, and ideas from statistical physics are widely applied across complexity science in many different ways (as explained in Chapter 4).

Physics is now a vast subject, and only a small part of it is concerned with the most fundamental theories of strings and quantum gravity. Most physicists work with models of matter and radiation that describe approximations or lower energy limits of more fundamental physics. One of the most important things we have learned from the history of science is that new, more empirically accurate theories can be related to the old ones they succeed and the latter retain some validity; for example, ray optics is an approximation to wave optics, and Newton's equations are low energy limits of the equations of relativity and quantum theory. This means that classical physics has not been replaced for many applications; for example, fluid mechanics is used to describe what appear to be continuous media such as flowing water, even though water is composed of molecules and is far from homogeneous at the microscopic scale. Physics is largely like chemistry and the rest of science in being about emergent phenomena (in the sense discussed in Chapter 1 and further discussed below). In what follows, it is explained how even the physics and chemistry of nonliving systems exemplify the truisms of complexity science stated in Chapter 1.

As noted in Chapter 1, the starting point of complexity science is the fact that some of the behaviour of large collections can be novel, in the sense that the parts on their own, or in small numbers with small numbers of interactions, do not display it. Emergence is surprising because what will happen cannot be anticipated by thinking about the behaviour of isolated individuals or collections involving only small numbers of interactions among individuals. Most of physics is about the entities, properties and processes that spontaneously arise when there are very many interactions among the parts of a composite system. There are many forms that this emergence can take. The physical world exhibits rich forms of structure at many different length and time scales, and different physical theories describe different kinds of emergent behaviour. For example, as discussed in Chapter 1, a large collection of molecules forms a gas that has emergent properties of pressure and temperature. There are laws that relate these quantities, and whole systems can be described using them much more economically than by describing each of their parts.

In practice, describing the individual behaviour of all the particles of a gas is not feasible, but it is possible to describe their statistical properties – for example, the way they are distributed across states with different energies given the overall temperature of the gas. Statistical mechanics relates the average motions and energies of the very many molecules in a gas to its thermodynamic behaviour, such as heat flows and changes of temperature. This shows how many molecules together can behave in interesting and novel ways. The study of how macroscopic properties of solids and liquids arise from the basic physics that describes their constituent parts is *condensed matter* physics. It is largely this field that Anderson drew upon in the celebrated paper 'More Is Different', mentioned at the start of Chapter 1.[1]

The above examples all involve a higher level of description that simplifies the underlying very complicated behaviour of a system. The theory of ideal gases is an extreme example, since it uses just three degrees of freedom – namely, pressure, volume and temperature – to describe a system that really has of the order of 10^{23} degrees of freedom. In general, emergence in matter is the emergence of order of various kinds. The ideal gas laws represent a kind of order in the space of possible states of a gas, ruling out combinations of values of the macroscopic properties that violate them. They are laws of 'co-existence' that apply at any given moment of time and say what possibilities for different properties are mutually compatible. Another law of co-existence is the Pauli Exclusion Principle, which in its simplest form says that no two electrons can have all the same quantum numbers.[2]

Dynamical laws represent a different kind of order. Newton's second law says how a system acted on by a force changes over time. The Schrödinger equation in quantum mechanics is a dynamical law, as are Maxwell's equations that describe how changing electric and magnetic fields interact with charges and electric currents. There are many more dynamical laws describing emergent structure of various kinds. Most take the form of differential equations. As mentioned in Chapter 1, emergence often means there are lower-dimensional equations of motion that describe the dynamics of emergent properties, and the full extent of the dynamics of the system can be largely ignored. The emergent properties are often the ones we observe. For example, in the solar system, the myriad gravitational and other interactions of the very many particles in the planets and the sun give rise to the emergent order of elliptical orbits of the planets described by Kepler's three laws. This is because over large distances only the gravitational force is relevant,

---

[1]Of course, molecules themselves are emergent entities since atoms are composed of subatomic particles, and protons and neutrons are themselves composed of quark and gluon interactions. Ironically, 'particle physics' is all about the interaction of quantum fields that are continuous even though particles are not. (Also deterministic theories can be approximations to non-deterministic ones and vice versa.)

[2]The most general form of Pauli's law says that collections of indistinguishable fermions (particles of half-integer spin) must be in antisymmetrised states.

and because the sun is so much more massive than the planets, so their effects on each other can be ignored. This is another example of how simple laws can arise. The price that is paid for this simplicity is that we lose track of the detailed movements of the parts of the system and only know about their overall behaviour. Furthermore, the orbits of the planets are not exactly elliptical. The emergent simple laws are approximate.

The separation of the dynamics of a system into the relatively slow and fast is one of the most important kinds of approximation in physics (and is very important for complexity science in general). For example, in a benzene molecule there are six carbon nuclei and many electrons, but the dynamics of the electrons can be treated as if it were separate from the dynamics of the nuclei. This is because electrons are very much less massive than protons and neutrons, so their motions and interactions are very much faster. The separation of the time scales of the dynamics is the basis of the Born-Oppenheimer approximation used in quantum chemistry. There are many examples of slow and fast dynamics within a single system in physics, and many other examples in complexity science, and the importance of this fact is one of the main themes of this book (see Chapter 4, Section 4.3).

Emergence also often takes the form of relatively sharp boundaries in time, space or both. A good example of a spatial boundary is the thermocline that occurs in bodies of water where the temperature drops off sharply so that there are two very different regions rather than a gradual range. A more complex example is that of snowflakes. They are untold billions of water molecules forming ice crystals of intricate geometry and of such variety that the probability of two identical ones forming in the entire history of the universe is negligibly small. There are about a hundred different features of large snowflakes, and the number of possible snowflake structures is of the order of 10^{100}. Yet they all share the feature of a star-shaped sharp boundary in space. Similarly, there are sharp changes in time. Dramatic changes in time in emergent structures can result from the numerous tiny changes in their parts. Consider a snow field on the side of a mountain. Friction between the individual flakes holds them at some average depth across the mountainside. However, gravity is pulling the snow downhill. A strong gust of wind or a step in the wrong direction can suddenly cause millions of tons of snow to move at over a hundred miles an hour.[3]

Sharp boundaries, and emergent entities and properties in general, are relative to scales of energy, space and time. At the very small scale, there is no sharp change in the properties of molecules in a body of water to corre-

---

[3]Such dramatic changes are the subject of the branch of mathematics called catastrophe theory.

spond to the thermocline, but at larger length scales, such as that relevant for a diver, the change is stark. Similarly, the dramatic changes studied in geology take place over time scales of a million or so years, but over very short time scales everything changes smoothly. In general, there are time scales over which things change and time scales over which they do not, and length scales over which they look the same and length scales over which they do not. It is remarkable that some things look the same at different length scales or even at all scales (as with a truly random walk or any fractal structure). Moreover, some phenomena are independent of scale. For example, the diffusion of molecules in solution and avalanches occur at all scales in sand and snow. ('Scale invariance' is an important concept in complexity science and is discussed more below and in Chapter 4.)

Gases have very little structure at any scale, but liquids like water are more interesting, and, when ice crystals form, there are correlations between the orientation of water molecules over distances about a billion times the distance between molecules. In this context, order, structure or form is inhomogeneity of some kind. For example, crystals exhibit order because the molecules in them are in geometrical structures like the tetrahedrons of table salt. In the solid state the molecules are not distributed randomly, but when salt is dissolved in water, the sodium and chlorine ions are not bonded and the molecules are distributed randomly. Similarly, the symmetry of snow crystals is actually a reduction in the complete translational and rotational symmetry of the positions of molecules in liquid water. The order of ice and snow crystals breaks the symmetry of liquid water. For this reason, the emergence of order in physics is often characterised as a process of 'symmetry breaking'. Much of Anderson's discussion of emergence is about how symmetry breaking is found at different levels in the different sciences. The idea of symmetry breaking is very general and can be applied to transformations of many kinds, including very abstract ones. For example, a network of people would be completely symmetrical if everyone were connected to everyone else so that swapping people did not change the structure of the network. This amount of symmetry is never present in a social network because of clustering (see Section 2.6 and Section 4.8 below).

Symmetry breaking in physics is associated with changes in energy and temperature and is described by the theories of statistical mechanics and thermodynamics. Gases turning into liquids and liquids into solids, and vice versa, are processes called 'phase transitions', and they have many special features. They give rise to relatively sharp changes in otherwise smooth variation. The transition of water to ice is like this. The structure changes radically as the temperature changes hardly at all, and 'latent heat' is emitted.

Correspondingly, when ice melts, there is a point where adding more heat does not increase its temperature but is consumed in breaking down structure. Above the transition temperature, there is no crystal lattice of any significant size, and below it there is one.[4] This abrupt change in properties is an emergent feature only of large collections. There is nothing in the basic physics of a small number of water molecules to suggest that phase transitions would exist.

Large collections of water molecules have a rich structure to their behaviour, and they also create form in the world. The combination of water and gravity carves the intricate fractal structure of mountain valleys and river deltas, producing persistent and large-scale forms, by processes such as freezing and thawing occurring many, many times. More is different partly because many iterations is different. Phase transitions of water are extremely important for life on Earth, but of course there are phase transitions between solid, liquid and gas in all other kinds of matter too. Phase transitions are emergent order in the dynamics of systems of many parts, and they may also produce structure in the systems with which they interact.[5]

The sharp boundary in the freezing of water comes where the density (the ratio of mass to volume) changes discontinuously. As mentioned above, this involves the emission of latent heat at this transition temperature. Phase transitions like this, involving latent heat and a discontinuity in the rate of change of some property of the system, are called 'first-order phase transitions'. Density is related to the thermodynamic quantity of free energy. In thermodynamics, the free energy is roughly a measure of the capacity of a system to do work and is related to other quantities, such as the entropy, the temperature, and the internal energy (more of all this in Chapter 4). The inverse of the rate of change of free energy with respect to pressure is proportional to density. In most first-order phase transitions, there is a discontinuity in the first derivative of the free energy with respect to some thermodynamic variable (in this case the temperature). However, the idea of phase transitions is much more general than this even in physics. In 'continuous' phase transitions in condensed matter physics, there is no discontinuous change in the obvious properties of the system. An example of such a continuous phase transition is magnetisation. As a lump of iron is heated, its magnetisation does not change sharply, and there is no latent heat released, but the rate of change of magnetisation with respect to temperature does change discontin-

---

[4]Liquid water contains very small clusters of molecules in crystalline form that form and melt incredibly rapidly.

[5]The production of structure by a system is how the environment can have a record of a system, other than just the myriad tiny changes in the rest of the world that the system has caused.

uously.[6] There are also continuous phase transitions involving very different phenomena, including Bose-Einstein condensation and superconductivity.

Phase transitions and broken symmetries can be described by an 'order parameter', which, in the case of the phase transitions of water, is the difference in the densities of the liquid and gas phases at some pressure and temperature. This quantity vanishes at temperatures higher than the transition temperature and gets bigger as the temperature gets lower. The order parameter for magnets is magnetisation, and there are others for superconductivity and other phases of matter. An important emergent feature is the critical point. With water, this is the temperature and pressure at which the properties of the gas and the liquid are the same and the transition between them is smooth. Other phase transitions also have critical points, and associated with them are 'critical phenomena' of various kinds. The study of phase transitions, critical points and the thermodynamic limit (when the number of particles becomes large) in statistical mechanics gave rise to a theoretical framework that has subsequently been applied widely beyond the original context of solids, liquids, gases, and their properties, like pressure and temperature. Near critical points, fluctuations occur at all scales. Hence, scale invariance is associated with phase transitions. A 'universality class' of models is one in which all the models behave in the same way near to phase transitions, and hence have the same 'critical exponents' describing this behaviour.

The critical exponent associated with a phase transition features in a 'power law', in which one quantity scales as another quantity taken to the power of the 'critical exponent'. An example is the degree of correlation between parts of a large system scaling as a power of the distances from each other. This occurs when a ferromagnet is cooled down below the Curie temperature and long-range correlations between the spins begin to form. These universal aspects to critical phenomena are found in many different contexts and in many very different kinds of process. Signatures of universal exponents and power laws are also found in many non-physical complex systems, ranging from the human brain to modern cities (more on power laws in Section 2.5 below and in Chapter 4).

Critical phenomena often involve the way that the degree of correlation changes between parts of a large system that are at different distances from each other. This occurs when ice melts and the long-range correlations re-

---

[6]Since it is the rate of change of a rate of change that is discontinuous, such phase transitions are often called 'second-order'. There are even higher order phase transitions. The differences between different kinds of phase transitions, and their definition, are subtle and highly technical (see Binney 1992).

ferred to above break down. Surprisingly, such long-range correlations are found in systems which are not driven externally towards a phase transition but are hovering by themselves close to a critical point. The observation was first made in a one-dimensional lattice of coupled maps (Keeler and Farmer 1986). It was later observed in a computer simulation of a two-dimensional cellular automaton called the 'sandpile model' (Bak et al. 1988). In this model, 'sand grains' fall onto grid cells, and, when too many pile up on one cell, sand pours onto neighbouring cells in avalanche-like bursts. The cellular automaton rules are probabilistic and very simple. The size distribution as a function of frequency of the avalanches obeys a power law. Both the coupled-maps model and the sandpile model exhibit criticality that is said to be 'self-organised', because no parameters have to be tuned in order to arrive at the critical point, as they do with, for example, the critical point of ferromagnets. Real piles of sand don't behave exactly like the sandpile model, but they do avalanche at some critical angle of slope depending on the shape of the grains.

When even very simple physical systems are driven by an energy or matter influx from the outside, forms of self-organised order can emerge. For example, when a tray of sand is shaken repeatedly, hexagons can emerge spontaneously. It is also possible to observe the spontaneous formation of waves in such a system (Ristow 2000). Self-organised criticality has been suggested to exist in many systems ranging from earthquakes to brain dynamics (Pruessner 2012). There is much debate about the nature of self-organised critical phenomena (Watkins et al. 2016). The famous Belousov-Zhabotinsky reaction (BZ reaction), described in 1958 by Belousov and in 1964 by Zhabotinsky (Madore and Freedman 1983), illustrates self-organisation in a nonliving system (Nicolis and Prigogine 1977). The reaction occurs in a mixture of various chemicals, including potassium bromate and various acids. While the system is driven by an influx of reagents, various geometric structures spontaneously emerge. There are chaotic oscillations in the pattern of spots of colour corresponding to the different chemicals that form and decay and reform. In this and similar cases, the equations describing the reaction are nonlinear. There are many examples in chemistry of molecular oscillators, and of the spontaneous generation of spatial structures, such as the so-called 'self assembly' of molecular nano-structures. These forms of emergence are found in systems that are 'open' in the sense that there is interaction between them and the environment. The Belousov-Zhabotinsky reaction and others like it occur only in open systems.

Closed systems tend towards equilibrium states of constant temperature, pressure, chemical concentrations and other system variables. In general, equilibrium states are states that do not change in some relevant respect over some relevant time scale. For example, a cup of coffee is said to be at thermal equilibrium when it has cooled to room temperature, because its temperature stops changing over a length scale of seconds and minutes. However, it is not really static, and if left for a few weeks, it would evaporate completely. There are many other kinds of equilibrium state depending on the kind of system – for example, the equilibrium state of a mechanical system in which none of the parts move and the equilibrium state of a combination of chemicals in which their concentrations remain constant. (The importance of the general notion of equilibrium is illustrated in Section 2.5 below.)[7] A crystal at room temperature is like a closed system in that all its molecules have found their most stable configuration and its temperature is constant. A gas in a container at equilibrium can be treated as a closed system, as can systems of condensed matter, even though they are really interacting through gravitation with the rest of the universe because the effects of it on them are so small.

An open, driven system does not reach a static equilibrium, but it may have dynamic equilibria. A dynamic equilibrium is a changing state but one in which the amount of change is constant and some of the system's parameters do not change over time. An example of a system in a dynamic equilibrium is a lake that receives water through precipitation and loses water through evaporation and infiltration into the underlying soil. While the water flows in and out of the lake, the amount of flow is constant, and water inflow and outflow balance each other out. Closed systems exhibit only the simplest kind of emergence. More interesting kinds of emergence occur in complex systems that are out of thermodynamic equilibrium or driven from outside in some way.

All the truisms of complexity science discussed in Chapter 1 are true of the study of nonliving matter and radiation. The hierarchical structure of even monatomic matter, from the subatomic scale to the macroscopic, shows that more is certainly different in many ways. Thermal and statistical physics is computational (and probabilistic of course), and many of the methods developed to study nonliving systems are now applied to living systems. In the ideal gas laws, there is a very simple kind of emergent order. Much more interesting kinds of order are found in the structure of phase transitions. Phase transitions and critical phenomena, in substances such as water, ferromagnets, superconductors and other exotic forms of matter, show that there are

---

[7]A very important kind of equilibrium in behaviour science is the Nash equilibrium of game theory, which is a situation in which two or more agents in a game have strategies such that none of the agents has anything to gain by changing them. The idea of thermodynamic equilibrium is not directly applicable to complex systems such as the economy and the World Wide Web.

different kinds of invariance and forms of universal behaviour among non-living physical systems. Self-organisation shows how complexity can arise from simplicity (see Prigogine 1978, 1980). The study of systems like river deltas involves multiple disciplines such as geology, hydrology and meteorology. Such systems exemplify how the difference between the system and the order produced by the system may be hard to disentangle, because there is feedback between the two as the existing pattern of erosion directs water into the channels already present, thereby deepening and reinforcing them. The next section considers the universe.

2.2   The Universe

All living complex systems depend on their histories. This is also true of many, if not all, nonliving things. It is true, for example, of the Grand Canyon, in which there is feedback between geology and erosion as water flows down channels created by previous flows of water. It is especially true of the universe as a whole, which has a history estimated to be around 13.8 billion years long. While it is not known how large the universe is, there is a region visible from Earth, called the 'observable universe', that is around 46 billion light-years in radius (one light-year is 9.46×10^{15} metres). Over very large length scales, above 300 million light-years or so, the standard model of cosmology regards the universe as homogeneous and isotropic, which is to say structureless. Closer up, things look very different. The extraordinary progress in astronomy in the last few hundred years has revealed that the universe has many different kinds of structure at many different scales and that the structure of the universe looks different at different times.

The standard model of cosmology posits an early universe that is both very hot and very small. No ordinary matter existed, only subatomic particles and radiation, and there were no stars or galaxies. As this early universe cooled down, the symmetries of high-energy physics broke, and the various forces separated. These changes are treated as phase transitions. In the process known as 'big-bang nucleosynthesis' hydrogen, helium, beryllium and lithium were generated. Once ordinary matter and forces had arisen, what happened over subsequent millennia was the result of nuclear forces and electromagnetism combining with the effects of gravity. The universe now consists of three generations of stars, each with successively higher concentrations of heavier elements formed by nucleosynthesis in stars. Type III stars are the oldest, and it is found that they contain very little metallic matter and no elements heavier than iron (which is number 26 in the periodic table). The production of heavy elements in stars only occurs late in their lifetimes, and these elements are distributed into the wider universe when stars explode in the form of supernovae. Later generations of stars form from the remnants of previous generations and contain more metals.

This ancestry of stars gave rise to the tremendous diversity we find on Earth. There are nearly 100 elements forming nearly 5,000 naturally occurring minerals, and, including synthetic compounds, there are about 60 million known chemicals. The structures we now see in the universe exist only because of its history, in which the remnants of stars that have exploded as supernovae provided the ingredients for the next generation of stars and the next. The smallest grain of sand on Earth would not exist without the birth and death of the stars that made it.

The basic components of the universe are galaxies consisting of stars and remnants of old galaxies, gas, dust and dark matter, all bound together by gravity, and often, like ours, orbiting a supermassive black hole. There can be anything from 10^9 to 10^{14} stars in a galaxy (10^{10} in ours), and there are 10^{11} galaxies in the observable universe. Galaxies come in three main types – namely, elliptical, spiral and irregular – although there are further variations, and galaxies are extraordinarily diverse in their properties. For example, around two-thirds of spiral galaxies, including our Milky Way, have a bar of stars across the middle. Our galaxy is a highly structured entity having four spiral arms, as well as regions of gas, and a number of orbiting smaller satellite galaxies. Galaxies attract each other gravitationally, and feedback between them affects their formation and evolution and their motions and collisions. An early use of computer simulation showed that collisions of galaxies, modelled very simply as Newtonian particles, could create spiral arms (Toomre and Toomre 1972). Galaxies collide over time scales of around a billion years, and there are very few stellar collisions, so at small length and time scales, there is no collision at all.

Galaxies form groups, clusters and superclusters, where the last of these contain tens of thousands of galaxies. These superclusters form part of even larger structures called 'filaments', one of which is 500 million light-years long, 200 million high and 15 million light-years deep. Correspondingly there are also voids in the universe in which there are hardly any galaxies, stars or matter. These can be on the scale of hundreds of millions of light-years. The whole observable universe obeys Hubble's law, according to which objects in deep extragalactic space are moving away from the Earth with a velocity that is proportional to their distance away.

Stars are classified according to the spectra of radiation they emit. The simplest feature is the colour corresponding to the largest peak of the spectrum. Colours range from blue through yellow and red to dark brown and are indicative of the temperature of the star, which can range from around 500 K to above 100,000 K, with most stars between 2,000 K and 50,000 K. Of course stars have different masses and can also be more or less bright; hence we have stars such as brown dwarfs and red supergiants. A graph can be plotted of the brightness of stars against their colours, and it is found that most stars are found in a band in this graph called the 'main sequence'. The properties of stars in the main sequence depend only on their mass. This is an excellent example of emergent order in the interconnected dynamics and properties of stars.

As well as stars, galaxies contain dust, gas and plasma that are emitted by stars when they become supernovae – that is, stellar explosions. (On long time scales supernovae are events since the explosions themselves last only for seconds and the expanding clouds of radiation they create only for days or weeks. On shorter time scales they are objects.) Interactions in galaxies, therefore, involve dust, gas, plasma, planets, stars, black holes, globular clusters, satellite galaxies and radiation, and, according to the standard model of cosmology, dark matter. As mentioned above, it is thought that feedback plays an important role in galaxy formation. The hot gas from supernovae creates so-called 'galactic winds', and these gases cool and then stars form out of them, starting the whole process again. There are also thought to be supermassive black holes at the centre of almost all galaxies. The space between stars, the interstellar medium, includes cosmic rays and magnetic fields, but it is almost entirely empty of matter. It contains only about a million atoms per cubic metre but has clouds of material within it that can be twice as dense, again illustrating the degree to which even 'empty' space is structured. At small length scales, there is no structure, because the density of astrophysical gas clouds is much less than that of what would be regarded as a vacuum on Earth.[8]

The above-mentioned dark matter is thought to hugely outweigh the visible matter with which we are familiar. Dark matter is posited to underlie the structure of the universe, forming filaments and vast halos on which the galaxies sit. Dark matter is necessary to fully explain the motions of matter within galaxies. However, it is unclear what properties dark matter has and whether or not it interacts only gravitationally or harbours new physics. Nothing is currently agreed about the properties of dark matter, except that it is massive and does not interact electromagnetically (and hence is dark).

One of the structures in the universe that has been studied most is our

---

[8]There are of the order of 10^{25} molecules per cubic metre in the Earth's atmosphere near the surface, about 10 million per cubic metre in a very good vacuum in a laboratory, but only about one per cubic metre in an interstellar gas (Chambers 2004).

solar system with its planets and their moons and the sun at its centre. The relative stability of the solar system and its various subsystems, also including asteroids and comets, is an emergent feature of gravitational mechanics. The solar system is subject to the electromagnetic effects of the sun's solar wind, which also interacts with the local interstellar medium. The solar wind is a complicated structure that has a rich array of interesting features. It is a stream of charged particles, including protons and electrons, which are ejected by the sun. It comes in two forms, slow and fast. The fast solar wind (which is less dense) is emitted by 'holes' in the magnetic field of the sun, and the slow solar wind is emitted largely from its equator. The solar wind forms large -scale spatial boundaries. The termination shock is the region in the solar system where the solar wind slows down sharply to subsonic speeds. The heliopause is the boundary of a kind of electromagnetic bubble around the sun and all the planets called the heliosphere. These, like thermoclines in the oceans, are examples of how at a large scale there can be relatively sharp boundaries that emerge from underlying smoothness.

The sun itself is remarkably structured and exhibits a striking form of emergent high-level dynamics. Every eleven years, for example, the polarisation of the sun's magnetic field changes. Another form of emergent universal structure is the solar system's disk shape, which is a result of its history as a dust cloud spinning around the sun. Disk shapes are common in the universe, because they are naturally created in spinning systems as a result of basic Newtonian dynamics. Within the solar system, the planets exhibit all manner of structure of their own, including their climates and weather and the interactions between them and their moons. For example, the periods of the orbits of three of Jupiter's moons are in the ratio 1:2:4. This is an example of 'emergent resonance', meaning that their gravitational effect on each other is regular and periodic, and in this case it is stable and self-correcting (Murray and Dermott 1999).

The universe contains a large number of components that interact with each other in a nonlinear way. There is a nesting of emergent structure on many spatial scales. Each galactic structure represents the history of the early universe and the symmetry breaking that gave rise to the fundamental forces and subatomic particles, as well as the more specific history of the galaxy's own formation. For example, the spiral structure of some galaxies records that they were formed by a collision between two parent galaxies, and the structure of the Earth records the production of heavy elements in stars before our sun had even formed. As pointed out above, every grain of sand has a very long history and requires many layers of emergence, each of which involves dynamics over different length and time scales and many open thermal systems. There are high-level laws in astrophysics concerning emergent entities and their properties, such as the main sequence of stars, the dynamics of galaxy interactions, and the universe as a whole.

The universe seems to have become more and more complicated since its infancy, and the most intricate structures in the universe of which we know (which are living systems) seem to be more complex than they have ever been. Even if this is so, we do not know if this increase in structure and complexity will continue or whether the universe will undo the hierarchy of complexity it has formed. There is also an apparent paradox in supposing that structure increases as the universe gets older, since it is usually thought that the initial state of the universe has very low entropy, which in physics is associated with high levels of order. While the entropy of the universe has increased monotonically with time, structure and complexity also seem to have monotonically increased with time over the history of the universe. Some of the solution of this paradox might be found in the fact that there is no agreed upon general definition of the entropy of the gravitational field and so much that is unknown about the universe. Also current thinking is that everything will end up in black holes, which will evaporate eventually, leaving the universe in a uniform state and wiping out its rich structure discussed above. How complexity relates to entropy and disorder and how, if at all, complexity can be quantified is clarified in the rest of this book, especially in Chapter 4.

2.3   The Climate System

The Earth's climate system is at the interface between the living and the non-living world. The Earth is 4.5 billion years old. During at least the first 2 billion years of its existence its atmosphere was almost completely devoid of oxygen (Bekker et al. 2004). The production of atmospheric oxygen was entirely driven by microbes, which facilitated and coevolved with geochemical cycles on a surface equipped with only a thin film of liquid water (Falkowski et al. 2008). The climate and life on Earth are a mutual product of each other.

It is important to distinguish between climate and weather. Weather is the condition at a certain point in time and space with respect to temperature, precipitation, wind and other meteorological parameters. Climate is, roughly speaking, weather averaged over a time scale of several decades. Climate is also the statistics associated with weather observations over time, such as frequency, persistence, and magnitude of hurricanes; droughts; and extreme temperatures.

The climate system of today is the sum of five major components and the interactions between them: the atmosphere (air), the hydrosphere (water), the cryosphere (ice), the lithosphere (land) and the biosphere (organisms). It is a thermodynamically open system driven by solar irradiation and also by plate tectonics and mantle dynamics. It changes over time under the influence of these external drivers, as well its own internal dynamics (IPCC 2013 2013c; Maslin 2013). It is also affected by the changing composition of the ocean and the atmosphere due to biochemical processes and, more recently, human impact (IPCC 2013 2013b).

Although they are open driven systems, the weather and climate systems exhibit macroscopic patterns which are stable in time, such as the eye at the centre of a storm or the annually recurring Atlantic hurricane season. Important large-scale stable patterns in the climate system include the three different zones of atmospheric convection, which are responsible for the Earth's desert belts and the high precipitation zones in-between. The convection zones originate in the constant variation in solar power influx with latitude, with time of day and with time of year. This variation is due to the curvature of the Earth's surface, the Earth's rotation around itself and around the sun, and the tilt of its axis relative to the plane of rotation around the sun. The result is a temperature imbalance between the equator and the North and South Poles that leads to physical forces directing air away from the equator.

The first of the three convection zones, situated between the equator and 30° latitude, is caused by hot air rising at the equator and drifting northwards and southward until about 30° north and south. Here, the air has lost most of its moisture and heat and drops back to the surface, where it forms the first desert belt of the Earth, of which the Sahara is part. At the surface, the air spreads out both towards the poles and back towards the equator. At the equator, the air heats up again to close the cycle of the first convection zone. The air which is not moving back to the equator but towards the poles meets the colder air coming from the poles at about 60° latitude north and south. Here, the warmer air is forced upwards by the colder, heavier polar air. This upward movement creates the first large precipitation zone and the boundary between the second and third atmospheric convection zones. These three convection zones constitute a very simple structure that arises out of the many complicated physical and chemical interactions between solar radiation, air, land surface and water.

Large-scale stable patterns are also found in the oceans. The greatest existing currents of water, the deep-water currents, are caused by the interaction between winds in the lower atmosphere and ocean surface water. Wind moves surface water via friction. The movement of surface water, as well as density differences between the layers, causes the movement of lower-lying water layers. This downward effect sets in motion a circular heat transport between the northern and southern hemispheres, the so-called thermohaline circulation, also known as the conveyor belt. The conveyor belt begins in the Gulf of Mexico, where strong solar irradiation causes water to evaporate, leaving the surface water very warm and very salty. The wind currents drive this water north, causing a surface current known as the Gulf Stream, which transports a hundred times more water than the Amazon River, the largest river in the world (Baringer and Larsen 2001). When this stream of salty warm surface water reaches the northern part of the Atlantic, around Iceland, it cools down and hence becomes denser. Here it sinks and is driven southward again, along the coast of Europe and Africa all the way to the South Pole. Around the coast of Antarctica, it mixes with the extremely cold deep water from the pole. From here, the current continues eastward to the south of Australia and eventually connects up with itself in the Gulf of Mexico. This closes the cycle.

The water circulation in the conveyor belt is responsible for the mild climate zone of Europe, which would otherwise be much colder given its location across the border of the second and third polar atmospheric convection zones. It is feared that human impact will lead to the shutdown of this thermohaline circulation (Rahmstorf 2000). However, interactions between the many different components and dynamics on different time scales make accurate and precise predictions very difficult (Clark et al. 2002). For example, the transport processes in the atmosphere and in the oceans are on different time scales to each other and also on time scales much slower than the variation of the driving force of solar irradiation.

Feedback is a ubiquitous driver of the state of the climate system. Two kinds of feedback can be observed: negative (stabilising) and positive (destabilising) feedback. An example of negative feedback is the temperature regulation of land surface. An increase in temperature due to sunlight leads to surface water evaporation, which causes low-level cloud formation. Clouds reflect a higher proportion of sunlight than clear air. Thus, cloud formation reduces the amount of sunlight reaching the surface, and, as a result, the surface temperature decreases again. An example of positive feedback is the melting of the polar snow cover. An increase in temperature leads to snow melting, which reveals the soil and rock underneath. These have a darker surface than ice and are less reflective. Hence, the cooling effect of the snow albedo is lost, and the further increase in temperature causes more snow to melt. Radiative processes involving absorption and leading to a reduction in snow cover generally happen on a very fast time scale. Hence, any compensating process preventing collapse would need to be on a similar time scale.

The time scales of processes involved are crucial for the development of feedback loops. The carbon cycle is a good example for the role of time scales in the climate system (IPCC 2013 2013b). Carbon exists in various molecular compositions in all of the five building blocks of the climate – the oceans, the atmosphere, land, ice and the biosphere. In the atmosphere, it is most abundant as CO_2, but atmospheric CO_2 represents only a tiny fraction of the carbon in the Earth system, the rest of which is tied up in reservoirs such as soil, the oceans and rocks. There is a constant turnover of carbon within and between these different reservoirs, the time scales of which vary from seconds, in animal respiration, to thousands of years or even, in the case of rocks, millions of years. The take-up of carbon by marine microorganisms happens on a time scale of hours. On an average, CO_2 molecules are exchanged between the atmosphere and the Earth surface every few years. The average time it takes for carbon to flow from the atmosphere to the oceans and back again is around one thousand years. Natural uptake of carbon into geological formations through chemical reactions takes a few hundred thousand years. This last cycle, involving sediment-bound carbon, is called the slow domain of carbon turnover. The fast domain of carbon turnover includes all the other more rapid processes such as photosynthesis and ocean absorption.

Until recently the slow domain of carbon turnover had very little interaction with the fast domain. Any small amount of carbon flux from the slow domain to the fast domain through volcanic eruptions and chemical weathering and erosion was quickly absorbed in the fast domain without affecting the stability of its cycles. However, since the beginning of the industrial era, set at 1750, the fast and the slow cycles have been strongly coupled through fossil fuel extraction from geological reservoirs. Between 1750 and 2011, the atmospheric concentration of CO_2 has increased by 40% (IPCC 2013 2013a). The processes in the fast domain cannot balance out such a massive release of carbon from the slow to the fast domain. Positive feedback loops have developed with severe consequences for the overall climate. The most well-known of these positive feedback loops is linked to the greenhouse gas effect. An increased amount of CO_2 and methane, another carbon molecule (CH_4), in the atmosphere leads to increased reabsorption of the infrared light emitted by the Earth's surface. This reabsorption traps heat inside the atmosphere and perturbs the energy balance between incoming solar radiation and outgoing terrestrial radiation.

Over the last 100 years, the Earth's global mean surface air temperature has increased by about 0.5°C. The increase of global mean surface temperature by the end of the twenty-first century (2081–2100) relative to 1986–2005 is predicted to be between 0.3° Celsius and 4.8° Celsius, depending on the details of the considered scenario (Pachauri et al. 2014).

The climate system has interconnected processes on many different length and time scales and both positive and negative feedback loops. It is the last complex system considered in this chapter which is, at least in many parts, nonliving. The next sections consider living complex systems and those constructed by living systems and the new kinds of emergence they exhibit. These new kinds of emergence include the maintenance of structure and function and adaptive behaviour of various kinds such as optimisation, prediction, decision making and ultimately consciousness and thought.

2.4 Eusocial Insects

Eusocial insects show levels of cooperation and organisation unrivalled in the animal kingdom, as laid out in the now classic book by Bert Hölldobler and Edward O. Wilson (2008). Eusociality is displayed in species of ants, bees, wasps, termites, and aphids (small sap-sucking insects). Eusociality is recognised by four main characteristics: insects live in groups, they cooperatively take care of juveniles so that individuals care for brood that is not their own, not all individuals get to reproduce, and generations overlap. Some of the advanced eusocial insect species have different morphologies for reproductive and non-reproductive individuals and even for different kinds of labour within the non-reproductives.

A colony of euosocial insects is a social organisation whose collective behaviour is comparable to a multicellular organism in which cells cooperate to keep the organism as a whole healthy, well-fed, and procreating, while the probability of individual cell death is much higher than that of the organism. Colonial insects cooperate to ensure the survival of the colony, while the survival rate of an individual is much lower than that of the colony as a whole. For this reason, colonies of eusocial insects are also called *superorganisms* (Hölldobler and Wilson 2008).

Eusociality is associated with adaptive group behaviour. Swarming is a form of adaptive group behaviour that arises from very simple individual behaviours, often involving the exchange of information between individuals. The next two subsections give examples of adaptive group behaviour in eusocial ant and bee species and explain how it is the result of spontaneous self-organisation.

2.4.1 Ant Colonies

The life of an ant colony begins when a queen mates with a fertile male, after which she digs herself a hole under ground and starts laying eggs. The eggs hatch after about a week. This first brood immediately begins to work. The newborns leave the nest to forage and feed the next brood. They also extend the nest by adding chambers to it. In the meantime, the queen keeps laying eggs, moving to deeper and deeper parts of the nest. She does not need to leave the nest ever again because she mates only once in her life. She stores sperm inside her body and lays up to several hundred eggs each day for years. In this way the colony grows until it reaches a typical average size that varies from species to species. Some have about 50 workers; others, such as those of leaf-cutter ants, grow to a size of several million. The survival of a newly founded colony is by no means guaranteed. Up to 90% of new harvester ant colonies die before they are two years old. Once they pass this age, though, they are often robust enough to survive for 20 years and longer (Gordon 2010).

What determines a colony's survival is its ability to grow quickly, because individual workers need to bump into other workers often to be stimulated to carry out their tasks, and this will happen only if the colony is large. Army ants, for example, are known for their huge swarm raids in pursuit of prey. With up to 200 000 virtually blind foragers, they form trail systems that are up to 20 metres wide and 100 metres long (Franks et al. 1991). An army of this size harvests prey of 40 grams and more each day. But if a small group of a few hundred ants accidentally gets isolated, it will go round in a circle until the ants die from starvation (Couzin and Franks 2003).

The life of an ant colony is based on communication. Any encounter between two ants is a form of communication, either by physical touch or by exchange of pheromones, which are chemicals secreted by the ants. For example, each worker carries a specific combination of molecules on the surface of its body. This molecular cocktail contains colony-specific pheromones and molecules resulting from an ant's work, such as molecules picked up on a forest floor indicating a forager ant. An ant's antenna acts as a 'nose', detecting the molecules by touching another ant's back. Thus, ants recognise the type of worker they encounter by that worker's 'scent'. Pheromones are also released by foragers at regular intervals on their way back from a food source to the nest. Ants searching for food recognise a nest mate's pheromone trail and follow it to the food source. They, too, will secrete pheromones on their way back to the nest and thus enhance the trail marking. This positive feedback loop results in hundreds of ants going back and forth between a food source and their nest, thus producing the ant trails that are so ubiquitous in nature.

The life of an ant colony is stochastic. Ants react to stimuli only some of the time, and the probability of a response increases with increasing stimulus. Patroller ants, for example, which are responsible for checking the safety of the nest entrance, are the first to leave the nest in the morning. Their departure is triggered by an increase in temperature at the nest entrance. However, there is no set temperature value at which patrollers leave. Rather, the higher the temperature is, the more likely a patroller is to leave. It is their return to the nest that triggers the forager ants to leave the nest and start their day's work. The chance for a forager ant to leave the nest increases with every encounter of a patroller. If there are too few patrollers coming into the nest, then the foragers do not leave. If this is the result of some danger outside the nest, then this is a feature keeping the colony alive. If it is the result of too small a colony and thus too few patrollers around in general, it is a destabilising feature. In a small colony, small fluctuations can be fatal.

An ant colony is a social system governed by division of labour: brood care, nest maintenance, patrolling and foraging. Task allocation is mostly determined by demand and opportunity in the environment, both of which are governed by stochastic rules of behaviour based on interactions between ants but also by genotypic effects (Schwander et al. 2010).

One environmental factor in task allocation is where inside the nest an ant happens to be. Young ants are more often working as brood carers simply because they have just emerged from the pupal case themselves and are therefore already at the brooding site ready to work. Older ants are more likely to work as foragers, because they have had more time to venture further away from the place where they hatched. Tasks are not set for life, and ants may switch tasks very often. For example, the detection of a plentiful food source leads to the recruitment of ants from other tasks to become foragers. Similarly, if a predator has eaten many patrollers, foragers will switch to work as patrollers the next day. If the nest is blocked by debris, forager ants will be recruited to maintenance until the nest entrance is cleared. It is the ants who happen to be nearby who are recruited first. Allocation of reproductive ability is different. Fertility is determined genetically, but the mechanism for deciding fertility of a larva is currently unknown (Klein et al. 2016).

Feedback is another mechanism that governs the behaviour of the colony. For example, given two food sources of different qualities, such as differences in distance to the nest, ants will quickly and with great likelihood settle on the source which is closer to the nest. With only a few ants out foraging, the bias for one source over another is proportional to the relative quality of the food sources, mathematically a linear relationship. Once the number of foragers passes a critical value, the preference for the more convenient source increases drastically. The bias for one source over another is now proportional to the power of the relative quality of the food sources, a nonlinear relationship, and it is one example among many considered by Deborah Gordon (2010).

The combination of stochasticity and feedback can lead to surprising phenomena. Army ants are known to build bridges across small gaps, such as between two sticks or a cut in a leaf, in order to get to a food source. Army ants are completely blind, and yet they build bridges consisting of hundreds of ants fully suspended across gaps tens of times larger than an individual ant. A few ants start to form a bridge at the tip of the obstacle, thus slightly shortening the path for the others. Once the beginning of a bridge is formed, the ants quickly move it closer and closer to the optimal connection between the two trail ends by making it longer and, depending on demand, wider. Position, length, and width eventually plateau out, and the final structure is surprisingly close to an optimal trade-off between length of path and number of ants diverted away from foraging to bridge building. Individual ants do not have the cognitive capacity to perform such an optimisation on their own. But by following local probabilistic rules, a colony collectively arrives at a close to optimal decision (Garnier et al. 2013; Reid et al. 2015).

Ants have memory. A forager ant, for example, is able to remember the number of times it has crossed a pheromone trail. The memory of an individual ant lasts for between seconds and days, depending on the species (Hölldobler and Wilson 2008). Forager ants waiting in the entrance of a nest for patrollers to pass by remember encounters for about ten seconds. Some forager ants can remember a repeatedly taken trail for several days. The collective memory of a colony as a whole, however, is generally much longer than the memory of an individual ant. While an individual red wood worker ant has a lifespan of about a year, red wood ant colonies remember foraging trails for several decades. The information is passed on by older ants guiding newly hatched ants to individual trees. Thus, the collective memory of a colony is orders of magnitude longer than that of individual ants.

Individual ants are capable of basic cognitive information processing which allows the colony as a whole to make its higher-level decisions. For example, an individual ant assesses the goodness of a potential new nest site by approximating its area. An ant visits a site several times, each time laying a pheromone trail while exploring the site. It recognises its own trail from the visit before and is able to 'count' the number of times it crosses its own path. The fewer crossings it counts, the bigger it perceives the site to be. The bigger the site appears to the ant, the more likely it is to advocate it to a nest mate by guiding it in a tandem walk to the new site. If enough ants consider the site a good choice, the colony as a whole will eventually make a collective decision to move there. This protocol for assessing the size of an area is similar to an algorithm devised in the eighteenth century, also called the Buffon's Needle problem (Mallon and Franks 2000). Since the ants engage in a collective choice between two or more alternatives, following the collecting and processing of information by individuals, this is a kind of 'quorum decision making'.

An ant colony is not a closed system. It may have intricate mutual dependencies with other living systems. As explained by Deborah Gordon (2010), some ant species build their nests on a plant and have adapted their behaviour to the specific physiology of the plant. An Amazonian ant species, for example, cuts the hairs of the host plant and binds them to a trap using a fungus which the ants cultivate in the nest. This trap provides the ants with insect prey and at the same time protects the plant from invaders. Other species prune the ground around their host plant to aid its growth, which in turn increases the available space for their nests. Such mutualism between insect species or between insect and plant species is very common. In some of these mutualistic relationships there is a third partner involved, so-called scale insects. In horticulture circles, scales are usually described as pests. But they can form a crucial link between an ant colony and its host plant. Scales live off the plant's sap, which they suck out with their stylet stuck into the plant. They excrete honeydew, which, if not removed, would eventually drown the scale. The ants protect the scales by feeding off their honeydew. They protect the plant by fending off insect predators. All the while the plant provides the ants with a home and feeds the scales, which in turn feed the ants. Labelling scales as pests is overlooking such circular dependency between different organisms which is abundant in the natural environment.

2.4.2   Honeybee Hives

When a young honeybee queen is ready to fly, she leaves her mother's nest to mate with male honeybee drones. She mates with about 10 – 20 of them before she founds her own hive or replaces her mother. A honeybee hive can grow to a size of up to 50,000 bees (Mattila and Seeley 2007). The sperm the queen has collected and keeps in her abdomen is enough for a lifetime. A queen signals that she is alive and well and that no new queen should be raised by spreading a particular chemical throughout the nest. A healthy queen lays about 1,500 eggs each summer day, and for each one she can decide whether to fertilise it. Fertilised larvae that are fed with specially nutritious food grow to become fertile females, queens; otherwise they grow into infertile female worker bees. Unfertilised eggs grow to become male drones. Drones mostly stay inside the nest, feeding on the honey produced by the female worker bees, or fly out to look for young queens with whom to mate. The drones' only task is to pass on the colony's genes, which encode the history of its biological development. Most of the eggs get fertilised because many more workers are needed than drones. The female worker bees are responsible for maintenance of the nest, brood care, foraging, and producing honeycombs and the honey to store therein (Seeley 2010).

A honeybee has individual information-processing capacity that goes beyond that of ants. An example of this is the dance that honey bees perform in front or on top of their nest mates to advertise a food source or an alternative nest site. A dance is a walk in half circles, during which the bees flap their wings and waggle their bodies, for from a few seconds to a few minutes. The 'waggle dance' has always drawn attention but was not understood until the 1940s. Karl von Frisch, an Austrian ethologist working at the University of Munich, discovered that the waggle dance was more than a random jig of excitement over a food source or nest site. Von Frisch realised that the dance was a symbolic code for the location and quality of a food source or nest site. The direction of the straight line of the half circle relative to the main axis of the hive encodes the direction of the location relative to the direction of the sun. The duration of the dance encodes the distance, with approximately one second of dance corresponding to a distance of 1,000 metres. The number of repeats of a dance signals the quality of the food source or nest site. Foraging bees that watch the dance and sometimes walk with the bee during the dance put this symbolic information into action by flying 'in a bee line' to the indicated location. The location information is accurate enough for them to find the site with very little searching. The remarkable discovery that bees have a *symbolic* language to communicate the location and quality of an object, as well as other contributions to the understanding of collective animal behaviour, earned von Frisch a shared Nobel Prize in Physiology or Medicine in 1973.

As with ants, task allocation in a beehive is stochastic and modulated by feedback (Seeley 2009). The workers in a hive decide which task to attend to depending on the signal they receive from their direct environment and from other worker bees. A forager returning to the hive with water, for example, needs to find a receiver bee to whom to unload the water. Any difficulty in finding a receiver bee signals an oversupply of water. The longer the search time for a receiver bee, the less likely the forager is to perform a waggle dance afterwards to advertise the water site. This stochastic feedback mechanism regulates the number of foragers going out, reflecting the need of the hive as a whole, while individual bees act only on local information. Since the decision is stochastic, it is possible that a foraging bee finds a receiver bee very quickly although the overall need for water is low. Such misleading events are very rare. The large size of a hive acts like an error correction for wrong signals. Bee colonies consist of up to tens of thousands of bees. The law of large numbers dictates that almost all bees receive the correct signal. The large size of the system makes its collective adaptive behaviour robust against the occasional non-beneficial actions of individual workers.

The temperature regulation of a beehive illustrates why hives are also called superorganisms, as mentioned at the start of this section. From late winter to early fall, during the brooding season of a hive, the core of the nest is kept at an almost constant temperature of around 35° Celsius. During this time the outside temperature can vary between -30° and +50° Celsius. The bees regulate this stable temperature level through a negative feedback mechanism. When the hive becomes too warm, the bees form ventilation tunnels to let in cooler air. When the hive becomes too cold, they heat up the air by increasing their individual metabolism, fuelled by the honey they have produced and stored. Another illustration of the superorganism is a hive's defense mechanism against fungi. When a hive's nest gets invaded by a fungus, the bees inside the hive start flapping their wings to generate body heat. This produces a fever state of the hive, heating up the nest just enough for the fungus to die while the bees stay unharmed. Temperature regulation in a beehive is an example of self-organised adaptive group behaviour.

Collective decision making is present in all eusocial insects. We saw several examples of it for ant colonies. Bees have several mechanisms in place to make a group choice, such as when to move to a new nest. When a beehive has grown too big for its nest site, it needs to move or split into two hives. Initially, scouter bees fly out to locate and inspect potential new nest sites. Back at the hive, they communicate the location and quality of a site to fellow scouter bees using the language of the waggle dance. The other scouts are likely to visit the site if it is advertised to be of high quality. If a scouter bee follows the directions to a site indicated by the dance of another bee and returns convinced, it is very likely to perform the same dance and to attract further bees to inspect the site. Low-quality sites are more likely to be ignored, and scouts will fly out to search for new candidate sites instead. This feedback mechanism brings more and more bees to a good-quality site. Swarms can locate, inspect, and report on a dozen sites in one afternoon alone. The method for deciding on a site is similar to quorum decision making exercised by ants. The bees sense how many other bees are at a candidate site at a given time. Once this number exceeds a threshold of about 20 – 30 bees, they consider this site to be chosen. They return to the nest, discourage other bees from dancing for alternative sites by head and thorax butting, and initiate the rest of the hive members to warm up their wings in preparation for the move. The head and thorax butting ensures that only one site is chosen and the decision process does not come to a standstill when two sites have similarly high quality. It is a negative feedback mechanism necessary to ensure that a decision is made.

More recently, scientists have found remarkable similarities between the collective decision making of social insects and decision processes in a primate brain (Seeley 2010). Social insect colonies and primate brains are clearly very different systems. They are composed of very different information-processing units, insects in one case and neurons in the other. A colony is an independent unit, while a brain needs to be part of a body to be alive. Their tasks are very different, too. Bees forage for food; brains initiate a thought or the movement of a limb. On the other hand, each system is composed of units very limited in their individual information-processing capacity while, as a whole, achieving a remarkable level of sophistication. A beehive's procedure of identifying and moving to a new nest site is similar to the neural process in a monkey's brain when it is initiating an eye movement. Researchers presented captive monkeys with a screen of moving dots, some moving left, others moving right (Britten et al. 1993). The monkeys were trained to decide in which direction the majority of dots were moving and to indicate their decision by looking in that direction. During this decision process the neural activity in the brain areas responsible for visual perception was recorded. The visual system has direction-sensitive neurons, responding to movements in one direction only, either left or right. The firing of directional neurons triggers an increase in the action potential of higher-level neurons in the visual system which effectively integrate the signal. Once the action potential of a higher-level neuron passes a threshold, it fires. A 'decision' is made in which direction the object is considered to move. If competing signals are present, it takes longer until a higher-level neuron exceeds the threshold since neurons that integrate opposite directions inhibit each other. This threshold-dependent decision with negative feedback prevents a standstill in the decision process and contradicting signals being sent to the motor area. It also guarantees that even small differences are perceived eventually. This neural process is, like the collective decision making of ants and bees, an example of quorum decision making. Like ants and bees, markets and economies can be said to make decisions. For example, markets decide prices. The next section is about markets and economies.

2.5   Markets and Economies

The first modern economist was the Scottish philosopher Adam Smith. In his famous book *An Inquiry into the Nature and Causes of the Wealth of Nations*, he described economic agents as selfish individuals who solely pursue their own happiness and are not interested in promoting the welfare of the society as a whole (Smith 1776). In his view, society nevertheless benefits from the sum of such individualistic actions. Smith described this beneficial effect as not only unintended but unpredictable. The behaviour of many individuals leads to novel consequences, and, he thought, agents seeking a beneficial outcome for all would in fact be less able to produce one. He wrote of an 'invisible hand' guiding the economy as a whole, and it is, according to Smith, due to this invisible hand that a greater good emerges. He was not implying the existence of actual yet invisible central control in a free market. Adam Smith's 'invisible hand' is a metaphor for emergent order without central control.

Smith's ideas are at the heart of classical economics and are often called on to show that unregulated markets are beneficial. While the benefits of free markets are much debated, free markets show many emergent regularities that arise without an overall controller. The law of supply and demand is a canonical example of an emergent market regularity. In a free-market economy the price of a product depends on the number of people who signal interest in buying it – its demand – and on the availability of the product – its supply. The law of supply states that if the price of a product goes up, its supply goes up, because more is produced or sourced. The law of demand states that if the price of a product goes up, its demand goes down, because fewer can afford it. The two most famous curves in economics are those of supply and demand, originally drawn by Alfred Marshall in his 1890 book 'Principles of Economics' (Marshall 1890). These two curves, plotted in the same graph against the price of a product, intersect at a single point. This is the point of price equilibrium, where the price a consumer is willing to pay equals the price at which a producer is willing to offer it.

The equilibration of prices is observed in daily economic life. For example, the price of fuel can vary between petrol stations nearby in location. Only a minority of consumers will go to the trouble of finding and visiting the cheapest petrol station when the difference in price is small. However, if the difference in price is large enough, more and more people will drive to the cheapest station. Other stations will adapt, and the prices will equilibrate again. Stable price equilibria and a balance of supply and demand are the result of a negative feedback loop between the decisions of producers and consumers. Recall from the discussion of matter and radiation that equilibrium states are states that do not change over relevant time scales. Prices in a market equilibrium still fluctuate but usually not over the time it takes to fill the tank.

The theoretical existence of a price equilibrium in a market does not guarantee that this equilibrium is actually reached. While market prices and demand move towards the point of price equilibrium, as time passes, advantages may get entrenched, and the equilibrium may never be reached. Thus, a theory of market equilibrium with a supply and demand law is compatible with a very unequal distribution of resources. Where this is very visible is in the distribution of wealth. The observation of unequal distributions of wealth goes back to the end of the nineteenth century. Historical data show that 10% of the populations in Britain and France owned 90% of the total wealth around the turn of the twentieth century (Piketty 2014, p. 368). The distribution of wealth was studied in detail by the Italian Vilfredo Pareto, who began his career as an engineer but later turned his attention to economics and social science (Pareto 1980). Pareto analysed taxation data and noted that the number of taxpayers in each income bracket rapidly decreased with increasing income. The distribution seemed to decay as a function of the income raised to some negative power, which is a very rapid decay. The inequality observed by Pareto is captured in the so-called 80–20 rule: 20% of individuals own 80% of the resources. This is now also called Pareto's law. Pareto's law has been found to hold for a large variety of statistical data. Eighty percent of links on the World Wide Web, for example, point to only 15% of web pages (Adamic and Huberman 2000; Barabási and Albert 1999). In 2016, 10% of the Earth's population owned 89% of global wealth (Credit Suisse Research Institute 2016). Other examples are the size and frequency of market crashes, the distribution of earthquakes, the number of species per mammal genus, the number of interactions per protein, the word frequencies in the English language, the intensity of solar flares, and the population sizes of cities (for a review, see Clauset et al. (2009) and references therein).

Pareto's law of income is an emergent nonlinearity resulting from a positive feedback loop between the amount of money owned and the amount of money earned – 'the rich get richer'. In a finance-dominated economy in particular, the likelihood of earning money is positively correlated with the money already owned because the return on financial capital is usually much higher than the growth rate of the economy as a whole (Piketty 2014).

Feedback mechanisms give rise to many different economic phenomena. Another example is the profit on investment known as 'return'. Companies invest in bigger production, because it increases their total profit and because it drives down the cost of production per item. The return per investment is large at first, but then gets smaller and smaller upon further investment until it vanishes and no further investment is profitable. This phenomenon of decreasing profit per additional investment is called 'diminishing returns'. Diminishing returns are a form of negative feedback. The system reaches an equilibrium state, which in this case is a steady supply and price of a product. Diminishing returns are also considered as a mechanism for maintaining competition. When one company dominates a market, it will eventually run into the wall of diminishing returns, which will allow other companies to catch up.

There are counter-instances to the phenomenon of diminishing returns in which further investment leads to increasing returns. Increasing returns made Google the dominant provider of online searches and other online services. There are many other examples of companies growing because if many people use their services, even more people will want to use their services, while they will incur relatively small additional cost in providing for more costumers. Rather than facing diminishing returns, these companies will grow faster upon further investment. This phenomenon is not exclusive to the digital economies in the twenty-first century. Alfred Marshall already remarked on it when he considered the possibility of monopolies. Both decreasing and increasing returns are the result of many interactions and are driven by feedback.

In the economic models discussed so far, the economic agents are very simple in their behaviour. They decide whether to buy or to sell a product based on its price or on the profit they would make. The modern version of Smith's selfish individual is a consumer who is influenced by only two factors in a decision of whether to buy a product: how much it costs and how much its possession is aligned with the consumer's preference. In microeconomics a measure of preference is called a utility function. The formal theory of utility has a long history (Stigler 1950a,b). John von Neumann and Oskar Morgenstern were the first to cardinalise utility (1947). They showed that, under certain conditions, rational agents make optimal decisions by always choosing to maximise their expected utility over all possible options for consumption.[9] The von Neumann–Morgenstern utility theorem forms the basis of what is now called expected utility theory, a pillar of standard microeconomic theory.

Since von Neumann and Morgenstern's work, more psychologically informed versions of economic agents have been developed. Most notably,

---

[9]The expected utility of an outcome is the utility of the outcome multiplied by the probability of that outcome.

in the 1970s, Daniel Kahneman, Amos Tversky and Vernon Smith introduced ideas associated with what is now known as 'behavioural economics', for which Kahnemann and Smith received the Nobel Prize in Economics in 2002 (Tversky died in 1996). Behavioural economics has been used very successfully to model a plethora of phenomena that can be understood from a psychological point of view but not from the rational-agent point of view (Kahneman 2003). The importance of irrational agents in emerging market phenomena such as crashes had been underestimated until the arrival of behavioural economics.

Financial Economics

Some of the most rapid developments in economic theory took place in finance theory from the 1990s onwards. One of the dominant assumptions in economic modelling at the end of the twentieth century was that all traders are equally well informed and every trader has access to complete information. In this world of 'perfect information', traders who value stock options take all past events and all foreseeable future events into account. Since every trader has complete and equal information, a stock price quickly approaches an equilibrium price which reflects its 'true' value. It is an equilibrium price, because it does not change until new information becomes available. Financial markets which are run by rational traders with perfect information never deviate much from equilibrium. Only small fluctuations about the 'true' value can occur. In this perfect world, a stock price changes significantly only when new information becomes available. When that happens, information spreads, traders quickly agree on the new true value of a stock, and a new equilibrium is reached about which the stock price fluctuates negligibly. The efficient equilibration of a stock market due to 'perfect information' is known as the 'efficient market hypothesis' (Fama 1991). This hypothesis was developed by Eugene Fama in the 1970s and earned him a shared Nobel Prize in Economics in 2013, together with Lars Hansen and Robert Shiller. In an efficient market, the only reason for a major change in stock price, such as a crash, is a major piece of new information. Bubbles do not exist. The dot-com bubble, for example, which burst in 2000, could not be accounted for with efficient market theory. At the end of the twentieth century it was clear that, while markets are very efficient information processors, they do not equilibrate, because the time it takes a financial market to reach equilibrium is longer than the time between events disrupting the system.

Assuming efficient equilibration of prices has consequences for the predicted statistics of trade data. In an efficient market, stock prices randomly fluctuate about the equilibrium price. This makes it impossible to make a profit because randomly fluctuating prices are unpredictable. In mathematical terms, this means prices move as if they were random walkers. A random walk is a movement in a random direction for a random distance at discrete time steps. Stock prices that follow a random walk are independent of each other, and the amount of fluctuation is independent of the value of the stock. A deviation from this random movement of stock prices about their equilibrium value, in a world of rational traders with perfect information, is caused by external events alone. The internal dynamics of such a perfect market are such that any external event is immediately absorbed into a new equilibrium price. The financial system, according to this random walk hypothesis, is an open system with no further structure to its internal dynamics, and prices equilibrate on a much shorter time scale than that of the external events driving the system.

With the advent of computers and computerised trade, it became possible to generate and analyse large sets of data and thus to scrutinise some of the traditional assumptions in financial economics such as the efficient market and random walk hypotheses (Beinhocker 2006).[10] It was quickly found that real data did not match the statistics predicted by a random walk model. As early as 1963, in the dawn of the digital age but before the arrival of computerised trade, the mathematician Benoît Mandelbrot suggested that the statistics proved the random walk hypothesis incorrect (reprinted in Mandelbrot 2013). Mandelbrot observed that large movements in prices (i.e., crashes) are much more common than would be predicted from a random walk model. His work was largely ignored until decades later, when standard financial theory was in more trouble. In later work, Mandelbrot and Richard Hudson (2010) found that not only did financial data exhibit structure, but they also did so over multiple time scales, from minutes to months and years.

Beginning in the 1990s, some physicists began applying ideas from physics such as scaling theory and turbulence to finance and economics. Amongst the pioneers were Eugene Stanley, Jean-Philippe Bouchaud, Rosario Mantegna, Yi-Cheng Zhang, Doyne Farmer János Kertész and Imre Kondor. They and others realised the potential of these methodologies to explain empirical statistical regularities in prices, such as fat-tailed distributions (a family of distributions of which power-law distributions are a member) and to design models including agent-based models of financial markets and human agents' behaviour.[11] One result of these efforts was that the limitations of equilibrium models of financial markets became apparent. While they still had their

---

[10]A good account of the history of economics and complexity is given by Beinhocker (2006).

[11]For an early, comprehensive review, see Farmer (1999).

use, concepts from non-equilibrium physics or, as Farmer and Geanakoplos (2009) called it, 'the complex systems viewpoint', gained traction. Econophysics is now a recognised field with dedicated research centres and chairs at universities.

As we saw above in the discussion of matter and radiation, one fundamental property of matter is that it undergoes sudden and dramatic changes in structure and properties during a phase transition. If a system is already close to the point of a phase transition (close to its critical point), only a slight change in external conditions brings about an abrupt change in the properties of the system. Didier Sornette (2003) advanced the hypothesis that stock markets are similar to physical systems of matter. He showed that financial markets, as well as many other complex systems, carry the mathematical signatures of phase transitions (Sornette 2002). According to Sornette, stock market crashes are a sign of self-organised criticality in financial markets.

Crisis in an Economic System

The field of economics is divided into micro- and macroeconomics. The decisions made by agents and the prices that emerge from them are the subject matter of microeconomics. The agents might be individual human beings, but they might also be firms or other corporate entities. Macroeconomics studies whole economies and their emergent properties of growth, inflation and unemployment, as well as the effects of government policy. The two branches of economics are clearly linked, but their relationship is not fully understood (Ross 2014) and, thus, often ignored. An instructive case is the 2008 financial crisis, set off by the collapse of the Lehman Brothers bank in the United States. The crisis, according to a common charge, had not been foreseen by economists. Macroeconomists largely believed the economy was stable and that there was little risk of a crisis. Nonetheless, financial economists had predicted the crisis in mortgage markets and had published models of the mechanisms by which this crisis could be transmitted into the extra-financial economy (Holmstrom and Tirole 1997). The problem was that the financial and the macroeconomic models didn't inform one another, while, in the real world, the financial economy and the macroeconomy are not isolated from one another.

Whether the crisis was foreseeable or not, it is clear that the events were contrary to many of the assumptions in standard economic theory. In addition to the ones considered in the beginning of this section – equilibrium, linearity, rationality and perfect information – there was another standard assumption in finance which failed: that of uncorrelated risk. The assumption in standard financial models is that the risk of one investment is not correlated to the risk of another investment. In a system with nonlinear dependencies on overlapping time and length scales, this assumption is too simplistic. The almost collapse of much of global finance in 2008 showed that probabilities of default are not uncorrelated. One of the triggers of what would become a domino effect was the sudden downturn of house prices in the United States. This downturn left many people unable to repay their loans on overvalued houses in a collapsing housing market, and it led to more and more lenders defaulting.[12] As a result, insurers had to pay out on many insurance policies at the same time – policies on which they, too, defaulted as they were only partially backed up by available funds (known as liquidity). The highly inter-linked structure of the housing market with the rest of the economy had been hugely underestimated. A single perturbation of the system at one end – the default of one lender – could percolate through the system in an unpredicted way and with unprecedented speed.

The links between economic agents – banks, mortgage holders, investors – were due to financial products that were relatively new and had a small market share at the end of the twentieth century but grew rapidly thereafter. One such product is a collateralised debt obligation (CDO). Generally speaking, there is a low return on a low-risk investment and a high return on a high-risk investment. Since the risk associated with mortgage holders de-faulting is too high for some investors (say, pension funds), and too low for others (say, hedge funds) there was room for profit in designing a product with more than one level of risk. To this end, debt obligations were collected into bundles, called tranches, and each tranche had a different level of risk and return. Now any kind of investors, risk averse or risk affine, could buy shares in the tranche which was right for them. The less risky shares, in the so-called 'senior tranches', were guaranteed to be paid out first but had less return associated with them. The more risky shares, with higher return from the so-called 'equity tranches', were paid out only once the senior sharehold-ers had been paid. Since CDOs are constructed as bundles, the links between debtors and creditors are hidden. The trading of CDOs created an increas-ingly large layer of hidden links in the global financial network (Beinhocker 2006).

Another, then relatively new, financial product relevant to the financial crisis in 2008 is a so-called derivative. Derivatives are a form of insurance. Any creditor can take out insurance on a loan from a third party such as a large bank. Should the debtor default on the loan, the arising loss of invest-ment is compensated for by the insurer. Such an insurance is a credit default swap, or CDS. A CDS allows for zero-risk investments. The United States in-

---

[12] 'To default' means to fail to meet the obligations on a payment.

surance corporation AIG used to be a large provider of CDSs. In 2008, CDSs written by AIG amounted to $400 billion (Davidson 2008). Investment insurance is not regulated in the same way health or property insurances are. Because of missing regulations, until 2009 a bank could give out as many CDSs as it liked without actually holding enough assets (liquidity) to back up the CDSs it had written (Houman 2009). For exactly this reason AIG collapsed in 2008 (Paul 2008).

Derivatives generated an unprecedented profit for investors, and their market grew fast in the decade leading up to the financial crisis. With the increase in derivatives sold, an unprecedented number of hidden links were created between financial institutions. Warren Buffett saw the dangers of these financial instruments quite clearly. In a report from 2002, he notes, "derivatives are financial weapons of mass destruction, carrying dangers that, while now latent, are potentially lethal" (Buffett 2002). These hidden layers of the global financial network, created by new financial products, facilitated and accelerated the spread of default through the system during the financial crisis in 2008.

Financial Ecosystems

In the wake of the global financial crisis in 2008, standard models of the financial economy – systems in equilibrium, consisting of rational, completely informed agents acting in an uncorrelated way – came to be viewed with much more skepticism. New paradigms for financial economic models were needed. Modelling the financial economy as a complex network of financial institutions linked to one another through loans and insurances is one such paradigm which has gained much attention (Battiston et al. 2016). Andrew Haldane, former executive director of financial stability at the Bank of England, advocated the complex network view as a much needed new paradigm for financial economics (Haldane 2009). According to Haldane and many others, a complex network view of the economy allows the transfer of insights from other fields which have used complex network tools for much longer, such as ecology and epidemiology.

In ecology, network theory is now a standard tool to study species relationships. In an ecosystem network the nodes are the species, and the links are predator-prey relations – who eats whom. In a healthy ecosystem, there is a balance in the diversity of species and in the relative abundance of predator vs prey. A feature known to enhance the stability of an ecosystem is the presence of 'modularity' in its network of species (Stouffer and Bascompte 2011). In a modular network not every species is connected to every other species, but there are pockets of closely linked species, while each such pocket is only loosely linked to other pockets. Any perturbation in parts of the system, such as the extinction of a species, is thus prevented from disrupting the system as a whole. Modularity and diversity are factors that create stability (Page 2010). The stability of financial networks is suspected to depend on similar structural properties (Haldane and May 2011). For example, it is reasonable to assume that the extensive trade of derivatives broke down the modularity of the financial trading network, which is believed to be one factor in the global financial crisis. Diversity, too, decreased in the run-up to the financial crisis. Banks' balance sheets and risk management had become increasingly homogeneous (Aymanns et al. 2018).

According to Haldane, one lesson of the financial crisis of 2008 is that a modest event in a complex adaptive network under stress can lead to wide collateral damage (Haldane 2009). The inter-linkage of the network heightens the impact of shocks and crashes and keeps the system as a whole away from a stable equilibrium. Haldane concludes that financial economics could learn many lessons from the pandemic of the severe acute respiratory syndrome (SARS) in 2002. SARS is an infectious and deadly flu-like virus found in small mammals; it mutated, enabling it to infect humans. The speed and scale of the subsequent spread among the human population was enhanced by the global transportation network, which increasingly links major cities of the world as if they were neighbourhoods in a single city. The SARS pandemic began with the first known case of atypical pneumonia in Guangdong Province, China (World Health Organization 2013). A few months later the first deaths were reported, and the disease had spread by air travel via Hong Kong to Vietnam and Canada. Four months after the first reported abnormal case over 300 people had been infected worldwide and 10 had died. Four weeks later the number had risen to over 1,300 infected and 49 people dead world wide. While infection spread, the number of infected grew exponentially. Six months after the outbreak, a total of 7,761 probable cases in 28 countries had been reported, with 623 deaths. The SARS virus was identified and its RNA sequenced six months after the outbreak began (Marra et al. 2003). Due to international collaboration in prevention measures and laboratory studies, two months later the World Health Organization was able to declare the SARS pandemic contained.

It was intervention by governments that ended both the SARS pandemic of 2002 and the financial crisis of 2008. It is interesting that the containment of SARS was helped by the existence of another tightly linked network – the World Wide Web and recently established web-based systems trawling for unusual health events (Haymann 2013). The Web is the subject of the next section.

2.6　The World Wide Web

In the 1980s a group of scientists at CERN, the international high-energy physics laboratory in Geneva, was looking for a convenient way of sharing data. Some universities and research institutes in the United States had already built an infrastructure for sharing data, the Internet, by physically connecting a handful of computers and servers. Various protocols existed for data sharing, such as the File Transfer Protocol (ftp). The team of scientists at CERN, lead by Tim Berners-Lee, added a combination of a programming language (html), which turns text pages into interlinked hypertext, and a protocol (http), which interchanges these hypertext or web pages with servers. This allowed data stored elsewhere to be directly linked to and retrieved via the servers. The World Wide Web was invented. The servers still need to be connected up physically, but the web pages are connected only virtually by means of hypertext links. Thus, the World Wide Web is two networks in one, the physical structure of the Internet and the virtual structure of the hypertext links between web pages. In 1993 CERN decided to place this new technology in the public domain. This decision was probably crucial for the development of the web technology as we know it today, which is compatible across platforms and browsers and not fragmented into proprietary islands (see commentary in Berners-Lee et al. 2010).

The World Wide Web has grown enormously since its invention. In 1999, there were an estimated 1.3 billion web pages, of which roughly 800 million were publicly indexable (Lawrence and Giles 1999).[13] In 2017, the number of indexable pages on the web had reached 46 billion (Worldwidewebsize 2017). The remainder are web pages that are accessible only if the precise http address of the site or that of another web page linking to it is known – so-called non-indexable pages – as well as those in the 'dark' net, which are accessible only with specific software, configurations, or authorisation, often using non-standard protocols.

The Internet – the network of computers being physically linked through fiber optic cables or otherwise – has also grown massively over the last four decades. Estimates for the year 2000 differ between 3 million and 100 million servers (Internet Systems Consortium 2012; Lawrence and Giles 1999). One estimate for 2017 is 1 billion servers (Internet Systems Consortium 2012). An exact count is difficult because there is no registrar of servers. To estimate the number of servers, algorithms send artificial messages and record the computers they pass on their way to their destination. However,

---

[13] A web page is 'indexable' if it can be found by a search engine – i.e., if it is reachable by following links from major known websites.

estimations of the actual number of computers from these individual paths are marred with statistical errors. Mapping the Internet is similar to mapping a network of tunnels by sending through a robot, which can take only one tunnel at a time. The chances that it will miss tunnels and entire regions are considerable. For a reliable estimate many repeats are needed given the size and constantly changing structure of the Internet.

The total number of web pages, in the billions, is in stark contrast to the small number of hypertext links one has to follow to reach any given page from any other page. The average minimum number of links one has to follow to reach any page is called the *diameter* of a network. In 1999 the diameter of the World Wide Web was estimated at 19 (Albert et al. 1999), which is a very small number compared to the billion web pages existing at the time. The result of this small diameter has been termed the *small-world effect*. The diameter is so small because a very small collection of web pages has a very large number of pages linking to them (the number of links is called their *degree*). These high-degree pages are the hubs, and they form the highway, so to speak, of the World Wide Web. Any web page can be reached quickly by going via these hubs. There are fewer than one hundred domains with more than one million links from other root domains. A root domain is the name one needs to buy or register, such as google.com. At the time of writing, the top three root domains with the highest number of links were twitter.com, facebook.com, and blogger.com (Moz 2018). The majority of domains has fewer than 100 links. This is reminiscent of Pareto's law and the 80–20 rule (see Section 2.5 above). Mathematically speaking, a large part of the degree distribution of individual web pages approximately follows a so-called 'power law' (Adamic and Huberman 2000; Barabási and Albert 1999). A distribution follows a power law when the probability of an event (number of links) is inversely proportional to its size (number of domains with this number of links) raised to some power (there are more details about power laws in Section 4.6.3 of Chapter 4). Networks with a power-law degree distribution are also called *scale-free* networks. Although real-world degree distributions do only approximately decay as a power law, and only across parts of the distribution, they are often referred to as scale-free networks (Clauset et al. 2009). Such 'quasi-scale-free networks' are ubiquitous in technology as well as nature. Approximately scale-free degree distributions have been reported for metabolic networks (Jeong et al. 2000), protein interaction networks (Jeong et al. 2001), film actor collaboration networks (Barabási and Albert 1999), scientific collaborations networks (Newman 2003), and food webs (Dunne et al. 2002; Montoya and Solé 2002).

In addition to its hub structure, the World Wide Web is highly clustered.

Groups of web pages are often highly interlinked while only sparsely linked between groups (Albert et al. 1999). It has been shown that a scale-free network that also exhibits a high degree of clustering is indicative of a hierarchical organisation (Ravasz and Barabási 2003). Small clusters of nodes organise into increasingly larger groups of clusters of nodes. The small-world effect of the World Wide Web is due to its hierarchical hub-structured topology, which it has grown into over the years (Huberman and Adamic 1999).

The Internet is a physical network while the World Wide Web is a virtual network, yet they have the same topology as a hierarchically organised network of clusters with a quasi scale-free degree distribution (Faloutsos et al. 1999). This is striking since they have very different constraints. The Internet is a network in physical space, while the World Wide Web is not embedded in any spatial geometry. It costs much more money to build a server than to create a web page.

A consequence of the clustered structure of the Internet is robustness against the random failure of servers. The networked structure means that almost all servers are connected to more than one other server. Should a server fail, requests for a web page can be routed through other servers. At the same time the Internet is very vulnerable to targeted attacks. Robustness against random failure combined with vulnerability to targeted attacks is a feature of all networks with a quasi scale-free degree distribution (Albert et al. 2000).

A new global infrastructure is currently being built by embedding short-range mobile transceivers into more and more everyday devices such as mobile phones, watches, refrigerators and cars, enhancing communication between people and devices and enabling communication between devices themselves. This *Internet of Things* is interconnecting physical and virtual things into communication networks (International Telecommunication Union 2012). The minimum requirement for devices to be part of the Internet of Things is their support of processing and communication capabilities. While tagging objects with radio-frequency identification, for example – as is done for many food and other consumer items today – allows for the collection of data in real time, so-called 'smart' devices can do more than that; they can process information and trigger processes on themselves and on other devices. Generally speaking, physical 'things' in the Internet of Things are those capable of being identified, actuated and connected. Virtual 'things' are those capable of being stored, processed and accessed. A whole new virtual and physical infrastructure is being created, with its own data sharing protocols and physical networks. Applications of the Internet of Things today range from 'intelligent' transportation systems and 'smart' power grids to 'e-health' and the 'smart' home. In 2016, the number of Internet connected devices was estimated to be somewhere between 6 billion and 18 billion (Nordrum 2016). The Institute of Electrical and Electronics Engineers (IEEE) expects that by 2020, 250 million cars will be connected to the Internet of Things and the number of connected 'things' will have reached or exceeded 50 billion (IEEE Communications Society 2015). This number is of an order of magnitude similar to the size of some of the larger natural networks and not far from the almost 90 billion neurons in the human brain. However, the connectivity of the human brain is far denser than that of the Internet of Things, since neurons can have up to 15,000 connections with other neurons (more of these in the next section).

2.7   The Human Brain

The human brain has a mass of one to one and a half kilograms and is made up of neurons as well as other cellular matter. The exact number of neurons in the human brain is unknown, but a recent estimate is 86 billion (Azevedo et al. 2009), with each neuron having tens of thousands of connections (DeFelipe et al. 2002). The most striking fact about the human brain is its ability to produce consciousness, while of course a single neuron or a small collection of them produces nothing even resembling an idea or a thought. It is undisputed that the number of neurons and the way they are connected is crucial for the cognitive capabilities of any organism. Estimates for the number of neurons in the common dog are around half a billion (Jardim-Messeder et al. 2017). The octopus vulgaris has about 500 million neurons, of which two-thirds are located in the arm nervous system (Young 1963). The honeybee, which is amongst the most sophisticated insects, has about 1 million neurons (Menzel and Giurfa 2001), and the primitive worm C. Elegans has 302 neurons (White et al. 1986). The basic chemical and electrical signalling mechanisms of all these animals are identical. In the evolutionary history of the brain neurons have diversified and become more sophisticated, but some of the neurons in the human brain are largely the same as those in the brain of reptiles, and some of the most basic mechanisms are found in even more primitive creatures (Anctil 2015).

The first scientist to develop a method for recording images of neurons was the Italian anatomist Camillo Golgi, around 1900. Santiago Ramón y Cajal, a Spanish anatomist and often called the father of neuroscience, discovered Golgi's technique, improved on it and developed his theory of the brain as a collection of different kinds of cells (Finger 2005). This would later be called 'the neuron doctrine', and it was opposed to the prevailing view at the time, the reticular theory, according to which the brain is a continuous mass with no distinguishable components or modules.

Ramón y Cajal discovered the diversity of morphology of brain cells (which exceeds that of all other cell types in the human body combined). His main insight was that a neuron has directionality with input and output corresponding to electric polarity. He proposed this as the basic mechanism of information processing. The idea that electrical signalling encodes information still underpins the current scientific model of the brain, though the role of chemical signalling is now recognised to be fundamental, and there may be as yet undiscovered mechanisms. In 1906, Ramón y Cajal and Golgi shared the Nobel prize for Physiology or Medicine in recognition of their work on the structure of the nervous system.

The brain is hierarchical in its physical as well as its functional structure (Nolte and Sundsten 2002). It originates at the top of the spinal cord, which extends from the brain stem down the backbone, and branches into nerves that extend throughout the body. The forebrain (which, as the name suggests, is largely though not entirely located in the anterior part of the brain), specifically the neocortex, is responsible for the most sophisticated, conscious behaviour. The hierarchical organisation is a result of the brain's evolutionary history. New structures are built on top of ancient ones so that there are vestiges of the reptile brain and the mammalian brain in the human brain, and the basic mechanisms by which they work are incorporated. The relative size of these regions has changed greatly in evolutionary history – for example, the hindbrain is very small in reptiles and much larger in mammals. The relationship between body mass and number of neurons in primates follows an approximate power law (Van Dongen 1998). The most recently evolved part of the forebrain consists of the basal ganglia and the enveloping two cerebral hemispheres. There are further structures such as the thalamus, hypothalamus, hippocampus, and the amygdala. In this way the brain carries the history of its development within its structure.

To a limited extent it is true to say that the forebrain controls the midbrain, which controls the hindbrain, but there is lots of influence in the other direction, so this is a big simplification. The hindbrain, which provides the connection between the spinal cord and the rest of the brain, is responsible for life-supporting activity such as heartbeat, breathing, and swallowing.

The pre-frontal cortex of the human brain (the most anterior part of the forebrain) has increased in size but more importantly become much more interconnected in the last 2 million years. More than any other brain structure, it is the cerebral cortex which makes us human. It contains the machinery for language, conscious perception, the control of voluntary movements, and intelligence. Modern imaging technology can visualise the locations in the brain where motor control, visual images, or even feelings such as regret are generated (see, for example, Coricelli et al. 2005).

Even during a lifetime, much of how the human brain works is plastic, in terms of normal development during both childhood and adulthood, as well as in the sense that regions can be often redeployed if others are damaged. For example, after a stroke other parts of the brain often take over a function that was temporarily lost because it was associated with a region of the brain that was damaged (Ward 2004). This form of adaptive behaviour contributes to the robustness of the brain's functionality. The brain as a whole, of course, constantly produces adaptive behaviour in the external world.

The information processing in the brain, and its input and output with the environment via bodily perception and action respectively, depends on *transduction*, which is the conversion of energy from one form into another. For example, the photoreceptor cells in our eyes convert the energy of the light that falls on them into electrical energy, whereas motor neurons, which end on muscles, turn electrical energy into mechanical energy.

Neuroscientists now have an understanding in fine detail of how neurons generate chemical and electrical signals (Purves et al. 2018). The basic structure of all neurons is that they have input and output regions. Information, in the form of electrical and chemical signals, flows from the former to the latter. In the output region of neurons the incoming electric signal activates a chemical (called a 'neurotransmitter') which binds to the connected neuron and thus translates an electrical signal into to a chemical signal at the synapse, which is the region of connection. There may be up to tens of thousands of synapses on the surface of one neuron (DeFelipe et al. 2002). Synapses are regulated by glial cells; however, little is known about how exactly they contribute to information processing in the brain (Azevedo et al. 2009).

A neuron fires when there is an action potential. The exact point in time when a neuron will fire is always subject to some amount of randomness, because action potentials are subject to thermal (and other forms of) noise. Interconnections between neurons, in the form of synapses, are developed and maintained by feedback. By firing, a neuron will excite or inhibit other neurons, and the effects of those excitations on other neurons can reach it on a time scale comparable to that of its own dynamics. Repeated firing strengthens the semi-permanent synaptic connections, establishing memory and learned behaviour, which are associated with groups of neurons that fire together in roughly the same pattern each time they are activated.

The brain is part of the wider nervous system, and in both there are many subsystems and processing tasks, often confined within them. For example, the photoreceptor cells in the retina are connected to other retinal neurons that process visual information prior to transmission to the brain. Information is conveyed to and from the brain by nerves but also by blood vessels, which contain hormones produced by both the brain and the body. The resulting signalling network has an unimaginable size: ninety billion neurons, each with several thousand incoming signals, generating a network of roughly a hundred trillion interconnections.

The brain shows more than anything that 'more is different', because of the richness of the emergence that results from its activity. Language, thought and mind are the ultimate cases of emergence. Consciousness, the first-person perspective, our sense of time passing and our awareness of our own consciousness are the most elusive objects of scientific understanding. However, the brain also produces an extraordinary range of highly sophis- ticated emergent features, most of which are sub- or unconscious. For ex- ample, reaching out to turn a door handle, face recognition and turning the head towards a sound are all very high-level processes that result from the firing of specialised neurons in specialised regions of the brain, and from the action of neurotransmitters, all interacting with the rest of the body and the environment.

More recently, a 'hierarchical predictive coding' approach to studying the brain has gained traction (Clark 2013). The assumption of the framework of hierarchical predictive coding is that the brain executes signal processing as well as signal prediction. The techniques of probability and computational learning theory are used to model the brain as a hierarchy of information- processing levels, each level receiving input signals from the level below and generating signals for the level above. One of the first examples of hierarchi- cal predictive coding was a model of the visual cortex (Rauss et al. 2011). At the lowest level of this model, photoreceptor cells collect visual input from the environment. This information is propagated upwards to a higher-level processing unit, where the signals are used to generate predictions of future signals from the lower-level neurons. These predictions are compared with the true future signals, and any mismatch is propagated further up through the hierarchy of neural processing levels. At these higher levels, a similar information processing takes place, and revised predictions of future input from lower levels are generated. This leads to feedback loops between the different levels of information processing. Some of the levels are initiating actions such as motor neurons initiating a hand movement. On the one hand, the use of predicted signals speeds up the reaction time. On the other hand, any prediction error will be costly for the organism. The assumption of the framework of hierarchical predictive coding is that the brain has evolved into an organism that minimises predictive error on all levels of its information-processing hierarchy.

The above illustrates how the brain has many similarities with other complex systems, such as numerosity, feedback, and probabilistic dynamics, leading to distributed decision making and a flexible division of labour. These similarities can go deep, as seen in the example of quorum decision making in bee colonies (see Section 2.4.2 above). As noted above, there are very many neurons and very many interactions between the parts of the brain; equally, there is a lot of diversity in neurons and in the structure of their connections. The study of the brain involves both computational and statistical modelling and many disciplines, including biochemistry, computer science and physics. The brain is the best example there is of a system that displays adaptive behaviour. The other complex systems of which individual people are parts, such as markets and social groups, and the ones we have collectively created, such as the economy and the World Wide Web exist only because of the complexity of the brain. The human brain and the collective products of human brains are the most complex systems known. The next chapter considers what this chapter as a whole teaches us about complex systems.

Chapter 3

Features of Complex Systems

This chapter uses the representative examples of complex systems discussed in the last chapter to arrive at a list of the distinctive features of complex systems. Chapter 1 explained the early history of complexity science in the 1970s. By the late 1990s the ideas and methods of complexity science had been developed and disseminated widely. There is a snapshot of the views of prominent practising complexity scientists in a special issue of *Science* on 'complex systems' (*Science* April 1999) devoted to a celebration of the new science. These ideas of some of the key figures are still representative of the field and are the starting point for our analysis.

1. "To us, complexity means that we have structure with variations." (Goldenfeld and Kadanoff 1999, p. 87)

2. "In one characterization, a complex system is one whose evolution is very sensitive to initial conditions or to small perturbations, one in which the number of independent interacting components is large, or one in which there are multiple pathways by which the system can evolve. Analytical descriptions of such systems typically require non-linear differential equations. A second characterization is more informal; that is, the system is 'complicated' by some subjective judgement and is not amenable to exact description, analytical or otherwise." (Whitesides and Ismagilov 1999, p. 89)

3. "In a general sense, the adjective 'complex' describes a system or component that by design or function or both is difficult to understand and verify. ...complexity is determined by such factors as the number of components and the intricacy of the interfaces between them, the number and intricacy of conditional branches, the degree of nesting, and the types of data structures." (Weng et al. 1999, p. 92)

4. "Complexity theory indicates that large populations of units can self-organize into aggregations that generate pattern, store information, and engage in collective decision-making." (Parrish and Edelstein-Keshet 1999, p. 99)

5. "Complexity in natural landform patterns is a manifestation of two key characteristics. Natural patterns form from processes that are nonlinear, those that modify the properties of the environment in which they operate or that are strongly coupled; and natural patterns form in systems that are open, driven from equilibrium by the exchange of energy, momentum, material, or information across their boundaries." (Werner 1999, p. 102)

6. "A complex system is literally one in which there are multiple interactions between many different components." (Rind 1999, p. 105)

7. "Common to all studies on complexity are systems with multiple elements adapting or reacting to the pattern these elements create." (Arthur 1999, p. 107)

8. "In recent years the scientific community has coined the rubric 'complex system' to describe phenomena, structure, aggregates, organisms, or problems that share some common theme: (i) They are inherently complicated or intricate ...; (ii) they are rarely completely deterministic; (iii) mathematical models of the system are usually complex and involve non-linear, ill-posed, or chaotic behaviour; (iv) the systems are predisposed to unexpected outcomes (so-called emergent behaviour)." (Foote 2007, p. 410)

Clearly, these people have very different things to say, not all of which are compatible. Some of these statements introduce ideas that are essential to complexity; others are too vague or otherwise unhelpful. For example, 1 may be true, but unless we restrict what we mean by 'structure' and 'variations' everything in the world will count as a complex system. Comment 2 asks us to choose between equating complexity science with chaos and nonlinear dynamics, which we argue below is a mistake; or equating complexity with having a lot of components, which is too simplistic; or equating complexity with a system with different possible histories on the one hand, which is again too simplistic, and a completely subjective answer to our question on the other, which is tantamount to giving up.

However, together these statements teach us a lot, as follows: 2, 3 and 4 together show the importance of the ideas of multiplicity, nonlinearity and interaction and the computational notions of data structures, conditional branches and information processing that are central to complexity science, as discussed in Chapter 1 and Chapter 4. Comment 4 also introduces the important feature of collective decision making which is a part of at least some complex systems.[1] Comment 5 raises the important idea, introduced in Chapter 2, and discussed below, that complex systems are out of thermodynamic equilibrium. Comments 6 and 7 emphasise that complexity arises from many interactions and feedback among components, which is the first of the truisms of complexity science noted in Chapter 1 and which was encountered throughout Chapter 2 (it is the first feature discussed below). Comment 8 emphasises the idea of emergence already emphasised in the previous chapters and discussed further below.

The next section explores these features and others in more depth. We distinguish between 'conditions' and 'products', where the former produce the latter. Abstracting from the quotations above and reflecting on the examples in the previous chapter gives us the following list of features associated with complex systems.

1. Numerosity: complex systems involve many interactions among their components.

2. Disorder and Diversity: the interactions in a complex system are not coordinated or controlled centrally, and the components may differ.

3. Feedback: the interactions in complex systems are iterated so that there is feedback from previous interactions on a time scale relevant to the system's emergent dynamics.

4. Non-equilibrium: complex systems are out of thermodynamic equilibrium with the environment and are often driven by something external.

The interesting thing about complex systems is that these conditions can give rise to the following products.

5. Spontaneous order and self-organisation: complex systems exhibit structure and order that arise out of the interactions among their parts.

6. Nonlinearity: complex systems exhibit nonlinear dependence on parameters or external drivers.

---

[1] Recall from the discussion at the end of Chapter 1 that some people think that complexity is confined to systems that display adaptive behaviour. We return to this issue below.

7. Robustness: the structure and function of complex systems is stable under relevant perturbations.

8. Nested structure and modularity: there may be multiple scales of structure, clustering and specialisation of function in complex systems.

9. History and Memory: complex systems often require a very long history to exist and often store information about history.

10. Adaptive behaviour: complex systems are often able to modify their behaviour depending on the state of the environment and the predictions they make about it.

Not all these features are present in all complex systems. Whenever any of the products are found in a system, they are the collective result of the conditions, but not all the products are found in all complex systems. Often products help produce other products – for example, memory is impossible without a degree of robustness, and adaptive behaviour can build nested structure and modularity. The following subsections consider each of them in turn in more detail and begin to assess whether each is necessary and/or sufficient for complexity on any or some conceptions of what complex systems are. Chapter 4 revisits these features and discusses them from a mathematical point of view.

3.1   Numerosity

Mere numerosity of interactions or parts can produce dramatic differences in behaviour (Anderson 1972). For example, as mentioned in Chapter 2, a hundred army ants put down on a flat surface will wander around until they die of exhaustion, but colonies of a million army ants exhibit 'collective intelligence' (Franks 1989). Most of the examples of complex systems discussed in Chapter 2 have a great many elements. In gases or systems of condensed matter there are very large numbers of particles. The numbers of molecules of gas in the atmosphere, transactions in the global economy, and connections in the Internet are all not just large but enormous. However, some complex systems have a relatively small number of components. For example, even a few organisms can engage in collective motion (swarming), and although a honeybee colony usually contains about ten thousand bees, the smallest colonies consist of only fifty or so bees. In the smallest animal brains there are about ten thousand neurons, while since close to ten thousand neurons die every day in a healthy human brain without any noticeable effect, ten thousand is not that many in the case of the human brain.

Although what is meant by a large number of interactions is vague, such vagueness is ubiquitous in science. For example, the Correspondence Principle states that quantum systems consisting of a large number of particles behave classically, without specifying what exactly 'large' means. Similarly, phase transitions and other critical phenomena arise only when systems are composed of many parts, where, again, no exact number can be specified just as there is no exact number of birds needed for a flock. In general, how much numerosity is needed depends on the system. However, even in those complex systems with relatively few components there are many interactions among them to generate the relevant complex behaviour, and in those with many components, like the brain, the huge numbers of interactions are also crucial to complex behaviour.

Interaction is the exchange of energy, matter or information (which always involves the exchange of energy or matter). The mediating mechanism can be forces, collision or communication. Without interaction, a system merely forms a 'soup' of parts that are independent and have no means of forming patterns or establishing order. Note that interaction needs to be direct, not via a third party or a common cause. For example, there are correlations between the pixels on a screen in a video game of tennis, but the image of a ball is not really caused to move by the image of a racket hitting it. There is no genuine interaction between the images at all. Thus, we require not merely probabilistic dependence but causal dependence.

Locality of interaction is not necessary. Interactions can be channelled through specialised communication and transportation systems that create long-range interactions between agents of a financial market; nerve cells transport chemical signals over long distances. It is important that the idea of interaction here is not just that of the physical dynamics, but of the *dependence* of the states of the elements on each other.

In some complex systems, the components are more or less identical, as with the individuals in a colony of social insects and physical systems composed of the same kind of molecules. Their collective behaviour gives rise to new kinds of law-like behaviour that would never be suspected in advance. The nature of the individual interactions is no different from the collective interactions, but when there are enough of them, very different behaviour results. The similarity of the parts is required for them to be subject to the same laws. However, while numerosity of interactions among parts is a necessary condition for any of the products of complex systems above, not all the parts or interactions have to be of the exact same kind, and indeed it may be important that they are not (we call this 'diversity', and it is discussed further below and in Chapter 4). For example, the atmosphere is composed of many similar parts, insofar as there are many units of small volumes of air that are similar and that interact to give rise to the weather. Yet, of course, oxygen, nitrogen and other gases are different in important ways in their interactions and properties, and these differences give rise to important features of the climate.

Often in complex systems, large ensembles of similar elements at one level form a higher-level structure that then interacts with other similar higher-level structures. For example, many cells make up a human body, many human bodies make up a group, and many groups make up a culture or society. The large ensembles of similar elements in these complex systems give rise to nested structure and modularity, as discussed below.

3.2 Disorder and Diversity

Liquid water is more disordered than ice, because the orientations and positions of the component molecules are much less correlated with each other in the liquid state than when they form ice crystals. Steam is even more disordered, because there are no intermolecular bonds at all, whereas there are lots of intermolecular bonds in liquids, which is why they keep a constant density and do not expand to fill any volume like gases do. Numerosity is not sufficient for even the most minimal kind of complexity that is displayed by any system in which there is a degree of spontaneous order and organisation, because the interactions must be disordered for the order and organisation to count as spontaneous. Spontaneous order emerges as a result of random interactions among parts, rather than being built into the system or controlled externally. For example, a gas displays spontaneous order during phase transitions. The order arises from the aggregate effect of many collisions between particles that are effectively probabilistically independent of each other because the gas has no spatial structure.

Correlations of different characteristic length and time scales for different systems are found throughout physics. For example, liquids exhibit correlations over medium time scales, whereas gases exhibit correlations only over very short time scales. On the other hand, solids exhibit correlations over both large time scales and large length scales. Where there is order, there is predictability in principle, though perhaps not in practice. Highly disordered systems can still have predictable statistical properties. Even gases have some order in the sense that their molecules are typically distributed among different energy states in a way that depends on the temperature of the whole gas. A completely disordered system would be one that is totally random in the sense of lacking any correlations between parts over space or time at all scales. The idea of complete disorder or complete order is an abstraction, and any real system has elements of both order and disorder.

As discussed in Section 2.1 of Chapter 2, order is inhomogeneity of some kind, and that means with reference to some set of properties, so disorder ought to mean homogeneity. Yet clearly in one sense a gas is much less homogeneous than a solid, because the positions of the molecules in a gas change much more over time. On the other hand, a gas at constant temperature is very homogeneous in the sense that the average distribution of its molecules stays the same, and, as pointed out in Chapter 2, there is a complete rotational and translational symmetry about it. Ideas of order and disorder are always applied to specific features, and care must be taken to define clearly what the relevant notions of order and disorder are.

In complex systems, disorder can exist at the lower level in terms of the stochasticity in the interactions between the parts, as well as at the higher level, in terms of structure which emerges from them and which is never perfect. Disorder in the interactions is a property of all the examples of complex systems discussed in the previous chapter. In many biological processes order at the higher level emerges from disorder at the lower level, and thermal fluctuations are necessary for them to take place. For example, protein binding to DNA is driven by the energy provided by thermal fluctuations (Sneppen and Zocchi 2005). The result is the transcription of DNA and the production of new proteins.

The fact that the order at the emergent level is never perfect and disorder remains has led to the idea that complexity lies between order and disorder (Waldrup 1992). In fact what lies between order and disorder is the structure of a complex system or the structure that is produced by a complex system (see Ladyman et al. 2013).

The absence of centralised control is another kind of disorder. Centralised control is when a special component controls an aspect of a system's behaviour. For example, thermostats control heating systems centrally, and herds of elephants exhibit coordinated motion because they follow the dominant female in the group. Many complex systems have to maintain control over some parameter, often their temperature, as with beehives and brains, but they are often not equipped with single control actions. Rather, in complex systems, order, organisation and control are to a greater or lesser extent, distributed and locally generated, and not centrally produced. With decentralised control no privileged individual is issuing commands, and yet parts still display coordinated behaviour, such as the collective motion of a flock or swarm or the temperature regulation of a hive. In the economy there are central banks and policymakers, but their degree of control of the relevant parameters is partial at best. Clearly, lack of central control is not sufficient for complexity because there may be a lack of control or order without this producing anything.

Diversity is when the components of a complex system vary, in their physical structure and perhaps also in their role. Quote 1 above takes such variation to be essential to complexity, and it is indeed found in many of the examples discussed in Chapter 2. Many sophisticated forms of adaptive behaviour would be impossible without diversity, since it facilitates the division of labour. For example, in brains there are neurons specialised to different tasks, and in financial markets there are specialised components such as banks. Diversity is not important to the flocking of birds, however. Diversity is neither necessary nor sufficient for any form of complexity, but it is important in many complex systems. Different forms of disorder and diversity are discussed further in Chapter 4.

3.3   Feedback

The interactions in complex systems are iterated so that there is feedback from previous interactions, in the sense that the parts of the system interact with others at later times depending on how they interacted with them at an earlier time. Feedback plays a crucial role in how disorder generates order, and it can give rise to stability and robustness. The presence of feedback in a system is not sufficient for complexity because it does not always give rise to some kind of higher-level order or stability.

In a sense feedback is present in all systems, if only because their parts gravitationally interact with each other and because those interactions depend on the results of earlier interactions, and so on. However, in complex systems the feedback is relevant to the dynamics of the system as a whole, and to be so it must take place on a similar time scale. In physics, most processes happen on well-separated time scales, which is the opposite of what happens in the domains of the other natural sciences. For example, feedback is irrelevant to the aggregate behaviour of gases but vital to chemical oscillators such as the BZ reaction. The climate illustrates well the role of time scales in feedback. The slow and fast carbon turnovers do not exhibit feedback; one happens on a time scale of thousands of years, the other on a time scale of minutes to days. The fast turnover finds a new dynamic equilibrium much more quickly than the relevant dynamics of the slow process.

In living systems and some systems that are derivatives of living systems, feedback can be what produces adaptive behaviour. All examples of such complex systems in Chapter 2 exhibit feedback, including flocks, hives, brains, markets and IT networks. Consider again the flock of birds. Each member of the group takes a course that depends on the proximity and bearing of the birds around it, but after it adjusts its course, its neighbours all change their flight plans in response in part to its trajectory; so when a bird comes to plan its next move, its neighbours' states now reflect in part its own earlier behaviour. In flocks and swarms feedback between the parts that start randomly distributed in their positions and movements result in correlations in their motions. In this way the order of the whole arises spontaneously. Feedback is what brings about coordinated behaviour and maintains the structure of the collective, despite the absence of a central control system.

Ants are able to undertake complex tasks such as building bridges or farms even though no individual ant has any idea what it is doing; left on its own it exhibits much simpler behaviour. Individual ants behave very differently when enough of them repeatedly interact with each other. The repeated interactions lead to feedback, and they bring about the building of nests and the rearing of offspring that is impossible for a small group with too few interactions.

Recall the distinction between positive and negative feedback made in Section 2.3 of Chapter 2. Negative feedback can give rise to stability, which can be a form of error correction, as in the brain, when feedback from sensory systems is used to correct motor systems, and as in the feedback between producers and consumers in a market economy. Feedback is often to be understood as feedback in a functional sense – for example, feedback of information. Feedback as error correction is often used in control systems, the paradigm of which is the Watt steam regulator, where the speed of rotation of the device interacts in a feedback loop with the steam engine to control the speed of the engine. However, this is central control as opposed to distributed feedback because there is a single component that has a special role. In sum, feedback is a necessary but not sufficient condition for any kind of complexity.

3.4 Non-Equilibrium

In Section 2.1 of Chapter 2 the thermal equilibrium of a system is defined as the state in which the thermodynamic properties of its macrostate are effectively unchanging. This is the macrostate that is reached after some time, whatever the exact initial microstate. The free energy of a system is its capacity to do work, and this is minimal at thermal equilibrium, while the entropy is maximal (given the value of conserved quantities such as the total energy). The thermodynamic equilibrium states of gases and condensed matter systems and their phase transitions have a rich structure. Apart from isolated samples of gases and condensed matter systems, the other examples of complex systems discussed in Section 2.1 are thermodynamically open.

An open thermodynamic system is a system with a net influx of energy or matter, which can be thought of as a free-energy gradient (either within a system or between connected systems). Open systems may be externally 'driven', as with a pile of sand being driven by the adding of more sand and the BZ reaction being driven by a flow of the reagents. The amount of order of some kind generated by complex systems can often be related to the steepness of the gradient of some quantity. For example, minerals are created in the Earth's crust by processes involving great heat flows and changes of pressure. More kinds of minerals are produced where there is greater variation in temperature and/or the concentration of chemicals (Hazen et al. 2008). Both of these examples can be associated with a steep gradient of the free energy of the system with respect to time. The dominant driving force of the Earth's climate is solar radiation

Systems with a net influx and outflux of matter or energy can be in a dynamic equilibrium (like the lake of Section 2.1). A system is said to be in 'dynamic equilibrium' or 'steady state' if some aspect of its behaviour or state does not change significantly over time – for example, the ratio of types of elements and their interactions. Many biologists think about adaptive behaviour as the maintaining of dynamic equilibrium in a thermodynamically open system. The maintenance of the steady internal state of an organism is called 'homeostasis'. Homeostasis more generally is a form of dynamic equilibrium in which a system that is changing is nonetheless stable in some respect. The free energy is important in living systems, because maintenance of structure is associated with the consumption of free energy (well adapted mechanisms minimise the amount of free energy needed). Living systems are open systems and maintain themselves out of thermal equilibrium by consuming food or, in the case of plants, by photosynthesis.

The notion of dynamic equilibrium applies to systems whose physical parameters, such as temperature and energy, are irrelevant to their dynamics, because of the way in which they are driven. For example, the temperature and energy of a bank's physical state are irrelevant to how it functions as part of the economy. A financial market is an example of a system that can be in a state of dynamic equilibrium and for which a temperature distribution is not defined. Since many complex systems are not physical systems, and hence have no specified temperature or free energy, the concept of thermodynamic equilibrium as such is not applicable to them. However, because of the tight link between thermodynamics and statistical mechanics and probability theory (Jaynes 1957), generalisations of temperature and thermodynamic entropy are used – for example, for the equilibrium analysis of complex networks (Albert and Barabási 2002).

The BZ reaction, discussed in Section 2.1 of Chapter 2, forms a type of 'Turing pattern' (Turing 1952). Alan Turing proposed a reaction-diffusion model to reproduce how patterns such as stripes can form in an embryo. It has since been shown that the original model by Turing could explain most biological self-regulated pattern formation (Meinhardt 1982). In these examples of pattern formation after a period of time, a steady state is reached in which the system maintains a pattern of colours, such as spirals, stripes or spots. During the steady state, chemical reactions are taking place, while the overall ratio of visible colours and hence molecular constituents remains approximately constant.

A system is in chemical equilibrium if the concentration of reactants is constant and if no external source supplies energy or matter to the system. In a state of chemical equilibrium, only reversible reactions are taking place, and no net energy is added to the system. The BZ reaction is not in chemical equilibrium because the concentration of the reagents changes over time. This notion of equilibrium used in chemical reactions is an instance of the more general notion of the dynamic equilibrium of a driven system that is widely applied to complex systems (see Section 4.4).

Numerosity and disorder and diversity are found in an isolated gas at thermal equilibrium, but as noted above, being thermodynamically open is necessary for the most minimal kind of complexity displayed by matter. In general, physical complex systems are out of thermodynamic equilibrium with the environment and are often driven by something external. These conditions, together with feedback between the parts of the system, produce the other features of complex systems (spontaneous order and organisation, non-linearity, robustness, history and memory, nested structure and modularity, and adaptive behaviour) that make them of such importance and interest.

3.5   Interlude: Emergence

There is no conception of complexity or complex systems that does not involve emergence. As discussed in the brief history in Chapter 1, the idea of emergence was central to the discussions that founded the Santa Fe Institute. The problem is in saying exactly what emergence is over and above the minimal idea of novelty in the sense of the whole behaving in ways that the parts in isolation do not.[2] As explained in Section 2.1, almost all of physics and all of chemistry deal with emergent properties in this sense, as exemplified by all ordinary material substances. For example, water is a transparent liquid that forms ice when frozen and steam when boiled, but none of these properties or processes can be applied to a water molecule or the atomic constituents of water – namely, hydrogen and oxygen atoms – which are themselves emergent from subatomic particles and fields. Similarly, a gas exhibits the properties of pressure and temperature and the gas laws relating them. The kinetic theory of gases, according to which heat is molecular motion and pressure is due to molecular collisions, connects the emergent properties of the whole with the properties of the parts.

Different kinds of emergence exemplified by the examples of the last chapter are listed below.

- The emergence of structure at different scales

  Ultimately as far as we know the physical structures at different scales in nature emerge from fundamental particles and fields. This is the kind of emergence exemplified by atoms and molecules, crystals and minerals, liquids and gases, geological features and the atmosphere of planets, and solar systems and galaxies. Sometimes the structure is highly ordered physical structure in space – for example, in the lattice of salt crystals. Sometimes the structure is in the way the system changes and in its dynamics, as in the BZ reaction (see Section 2.1).

  As emphasised in Chapter 1 (truism 9), it is important to distinguish the structure and order of a system and the structure and order it produces – beavers and beaver dams, bees and honeycombs, and elephants and their graveyards. The last kind of order or structure may persist long after its production has stopped (human burial mounds are visible millennia later). Structures produced by complex systems are not in general complex systems themselves but may be studied as products of them.

- The emergence of dynamics, properties and laws

  The gas laws are emergent, and they relate properties of a whole gas that its parts do not have. One important kind of emergence is that there is a lower dimensional effective dynamics within a higher dimensional

---

[2]As Paul Humphreys says, "In the case of emergence, there are too many different uses of the term 'emergence' that are entrenched across various fields for a single comprehensive definition to be possible at this point in time or, perhaps, ever." (2016, p. 26). A detailed analysis of the current disagreements about emergence is provided in Wilson (2015).

state space. That is the case for a ball rolling on a table. The number of degrees of freedom can be massively reduced if the collection of particles constituting the ball is described as a single entity. Similarly, the approximately elliptical orbits of Kepler's laws of planetary motion emerge over time from the gravitational interaction between all the particles of the sun and the planets. Emergent properties and laws massively reduce the number of degrees of freedom needed to describe and predict the behaviour of the system.

However, there is a trade-off as the emergent laws do not say anything definite about the individual parts of the system. In general, descriptions of systems at a higher level say only probabilistic things about the lower-level entities that make them up, but this is often all that is needed. The alternative would be that, in practice, nothing could be said at all about the system dynamics because the calculations would be impossible even with supercomputers. Compare trying to predict what will happen next when a mouse and an owl catch sight of each other using the laws of fundamental physics applied to their parts with using behavioural biology.

As explained below, emergent dynamics or emergent laws may be non-linear even though the underlying system is linear. The higher-level dynamics may even appear random while the system is deterministic, as is the case with chaotic systems. Often there is emergent relative decoupling of dynamics at different time and length scales, as illustrated by the examples of the climate and the solar wind.

- The emergence of various kinds of invariance and forms of universal behaviour[3]

  Critical phenomena, phase transitions and scaling laws are another kind of emergent structure that can be found in very different physical systems, as discussed in Section 2.1. In complexity science critical phenomena show up in probabilistic descriptions of geological phenomena and in network descriptions of biological and social systems, as discussed in Chapter 4. Recall truism 5 from Chapter 1 which states that complex systems are often modelled as networks or information-processing systems. Doing so allows novel kinds of invariance and forms of universal behaviour to be studied.

---

[3]This is truism 6 from Chapter 1.

(1) Networks

The parts and interactions in complex systems may often be considered abstractly and independent of their physical nature. This makes complex systems well suited to being modelled as networks. Complex systems are composed of many components interacting repeatedly, but some components may interact more than others, and some may be well connected and others less so. A new kind of universality emerging from this point of view is a scale-free degree distribution (see Section 4.6). Many of the properties and products of complex systems discussed below can be found in networks. Network theory is discussed in more detail in Chapter 4 and in the Appendix.

(2) Computation/information-processing systems

Bold claims are sometimes made about how life, or even the universe itself, performs computations or maybe that somehow it is just computations or information (see, for example, Wolfram 2002). This may be wrong, but it is certainly true that it can be extremely useful to model systems as such. The multiple pathways of quote 2 at the beginning of this chapter are emergent at a higher level of description than the physical description of a complex system. They may be computational pathways or flows of information. The emergent level of description in which the system is considered as processing information is of particular importance in complexity science. Many measures of complexity use information theory, as discussed in Chapter 4.

Probability and statistics are the underlying tools for the study of networks and information, and the application of statistical mechanics to complex systems outside of physics, as explained in Chapter 4, relies on them.

The next sections consider various emergent features of complex systems that take different forms in the context of the different kinds of emergence above.

3.6   Order and Self-Organisation

Clearly a fundamental idea in complexity science is that of order and self-organisation in a system's structure or behaviour that arise spontaneously from the aggregate of a very large number of disordered and uncoordinated interactions between elements. Even simple condensed-matter systems and gases exhibit the minimal kind of order and self-organisation discussed in

Section 2.1. The dynamic equilibrium in the patterns generated by chemical reactions like the BZ reaction is a kind of spontaneous order. Spontaneous order is exploited by applications of complexity science such as methods of 'self assembly' in chemistry. The generation of order involves symmetry breaking. For example, when a crystal forms, the translational and rotational symmetry of the molecules in the liquid is broken. Self-organisation may mean collective motion and various kinds of organised behaviour, as discussed below.

It is far from easy to say what order and structure (as well as self-organisation) are. Notions that are related include symmetry, periodicity, determinism and pattern. As pointed out in Section 3.2 above, the notion of order may mean so many things that it must be carefully qualified if it is to be of any analytical use in a theory of complex systems. One of the most difficult issues is how order and self-organisation in complex systems relate to the information content of states and dynamics construed as information processing. The problem is that the interpretation of states and processes as involving information may be argued to be of purely heuristic value and based on observer-relative notions of information being projected onto the physical world. Chapter 4 returns to the role of information theory, and the idea of order is made mathematically precise in Section 4.5 of Chapter 4.

It is a necessary condition for a complex system that it exhibit some kind of spontaneous order and self-organisation. Of course, order and self-organisation must be relatively stable to be worth identifying. As with the notion of dynamic equilibrium, what counts as 'stable' depends on the relevant time and length scales (see the discussion of robustness below). All the other products of complex systems require some kind of order or organisation.

3.7 Nonlinearity

In the popular and philosophical literature on complex systems a lot of heat and little light is often generated by talk of linearity and nonlinearity. For example, in his widely cited book, Klaus Mainzer claims that "linear thinking and the belief that the whole is only the sum of its parts are evidently obsolete" (1994, p. 1). It is not explained what is meant by 'linear thinking' nor what nonlinearity has to do with the denial of reductionism. Unfortunately the discussion of complexity abounds with non sequiturs involving nonlinearity. However, nonlinearity must be considered an important part of the theory of complexity if there is to be one, since certainly many complex systems are also systems that obey nonlinear equations such as power laws.

As explained in Chapter 1, the mathematical definition of linearity is that a system is linear if one can add any two solutions to the equations that describe it and obtain another, and multiply any solution by any factor and obtain another. Nonlinearity means that this 'superposition' principle does not apply. It may refer to dynamical equations or to any equations describing the relationship between variables. Nonlinearity is often considered to be necessary for complexity (as suggested by quotations 2, 5 and 8 at the start of this chapter). This claim is misleading. Complex networks, for example, which are much studied in complexity science, can be defined by matrices which are mathematically linear operations. There are also complex systems subject to game-theoretic and quantum dynamics, all of which are subject to linear dynamics. This shows that, in general, nonlinearity in the dynamics of the elements is not necessary for complexity, while nonlinearity on the level of the whole is often a feature of complex systems.

Nonlinearity is also not sufficient for a complex system (on any but the most all-encompassing conception), because, for example, a simple system consisting of a pendulum can be subject to nonlinear dynamics and exhibit chaotic behaviour, but it exhibits only the most minimal kind of emergence without self-organisation. Complexity is often linked with chaos, and as noted above, it may be conflated with it, but the behaviour of a chaotic system is indistinguishable from random behaviour. It is true that there are systems that exhibit complexity partly in virtue of being chaotic, but their complexity is something over and above their chaotic nature. Furthermore, since chaotic behaviour is a special feature of some deterministic systems, any dynamical system that is stochastic is by definition not chaotic, and yet complexity scientists study many such systems.[4]

So it seems that chaos and nonlinearity of the underlying dynamics are each neither necessary nor sufficient for complexity.[5] The fact that nonlinearity is a commonly mentioned feature of complex systems is an example of the conflation of having nonlinear dynamics and being complicated yet defined by simple rules (where 'complicated' means difficult to predict).[6]

However, there is a sense in which complex systems are never linear sums of their parts. The emergent behaviour of complex systems comes about because the parts interact so what the individuals do together is different from the sum of what each one of them does alone. For example,

---

[4]Note that chaos as in 'chaos theory' is always deterministic chaos.

[5]Robert MacKay (2008) argues for a definition of complexity as the study of systems with many interdependent components and excludes low-dimensional dynamical systems, and hence many chaotic systems.

[6]Arguably, unlike complexity, chaos can be simply defined as so-called strong mixing, meaning that the system's state is effectively independent of its earlier state (Werndl 2009).

individual ants will just wander around, but together they will behave very differently. Hence, nonlinearity may be both a property of the system's dynamics that contributes to its complex behaviour and it may also be a product of the complex system in the sense of being a form of emergent order – for example, power laws in statistical distributions of city sizes or income (see Section 2.5). Nonlinearity in some guise, either of dynamics or of emergent relations, is a very common feature of complex systems, and it is discussed much more in Chapter 4.

3.8   Robustness

Robustness is of some kind of structure, order, self-organisation, law or function. The most basic kind of emergence occurs only when the order that arises from interactions among parts at a lower level is robust. It should be noted of course that such robustness is only ever within a particular regime. The emergence of condensed matter requires a certain range of temperature, because at very high energies, molecules all break down and atoms lose their electrons and become plasma. At even higher energies there are no atoms only subatomic particles, and so on. In living systems, the conditions in which emergent structure is robust may be very limited. For example, the interactions among our neurons generate an emergent order of cognition, but if the brain is heated to about $5°$ Celsius above its normal temperature, it breaks down.

The order in complex systems is often robust because, being distributed and not centrally produced, it is stable under perturbations of the system. For example, the order observed in a flock of birds, despite the individual and erratic motions of its members, is stable in the sense that the buffeting of the system by the wind or the random elimination of some of the members of the flock does not destroy it. A centrally controlled system, on the other hand, is vulnerable to the malfunction of key components.

The term 'resilience' is sometimes used interchangeably with 'robustness', but it is better thought of as a kind of robustness - namely, the ability to recover from a perturbation on a time scale which is short compared to the system's lifetime. Resilience may be formulated in computational language as the ability of a system to correct errors in its structure, and this may be achieved by exploiting feedback. In communication error correction can be achieved by introducing a limited amount of redundancy. This redundancy is usually not explicit, such as a copy of the string or its parts, but more subtle – for instance, exploiting parity checking, which is more computationally intensive but also more efficient (the message is shorter than simple duplication) (Feynman 1998). In his influential account of complexity (discussed in Chapter 4) Charles Bennett (1991) specifically mentions error correction: "Irreversibility seems to facilitate complex behavior by giving noisy systems the generic ability to correct errors". For example, a living cell has the ability to repair itself (correct errors), as when a malfunctioning component is broken down and released into the surrounding medium. In the cell, a small perturbation cannot be allowed to spread to all the degrees of freedom. Hence, the cell has a one-way direction for this dispersal, errors within the cell are transported out, and possible sources of error outside the cell are kept out (assuming the errors are sufficiently small).

Many similar elements interact in a disorderly way in a gas, but it is not a complex system in any but the most trivial sense of obeying emergent laws. However, a system consisting of many similar elements that are interacting in a disordered way has the potential to form patterns or structures. An example is the Rayleigh-Bénard convection patterns that form in layers of liquids subjected to a temperature gradient, as well as the patterns made by the BZ reaction and other self-organising chemical phenomena. On an appropriate time scale the order is robust. This means that although the elements continue to interact in a disordered way at small length and time scales, the larger scale patterns and structures are preserved. A macroscopic level arises out of microscopic interaction, and it is relatively stable. This kind of robust order is a necessary condition for a system to be complex. A good example of robustness is the climatic structure of the Earth, where rough but relatively stable regularities and periodicities in the basic phenomena of wind velocity, temperature, pressure and humidity arise from an underlying nonlinear dynamics.

Some kind of robustness is necessary for any kind of order to exist and is therefore necessary for complexity. However, order may persist without being maintained by the kind of emergent processes born of stochastic behaviour that is characteristic of complex systems. On the other hand, in a completely random system perturbations do not affect the state. Hence, robustness is not sufficient for order (or any other products of complex systems).

There is obviously a trade-off between robustness and adaptation, because systems that are very robust adapt more slowly. A salt crystal is very robust until it is dropped on the floor and breaks to pieces. It can reform but only after being dissolved in water and being left until the water has evaporated. To conclude, robustness of some kind is a necessary component of every complex system and is discussed at length in Chapter 4.

3.9 Nested Structure and Modularity

Anderson (1972) argued for the importance of considering hierarchies of structure to understand complex systems. There is nested structure in the physical world from the subatomic through the atomic to the chemical, and ultimately to planets, stars, galaxies, clusters and superclusters. The Earth and its oceans and atmosphere exhibit a very rich nested structure, as does the solar system. In such complex systems there is structure, clustering and feedback at multiple scales.

In systems that perform functions there can be the division of function or labour and the emergence of levels of functional organisation. The division of labour in an ant colony is an example of functional organisation, in the sense that there are subsystems for subtasks, but this is true even though the worker ants are more or less identical. In many other cases functional organisation goes along with differentiated structure that is specialised to fulfill different functions, as in even simple cells. Multicellular organisms divide labour such as respiration, digestion and excretion, as well as cognitive capacities such as memory and perception, among highly specialised parts.

Cities, economies and many other social systems are systems composed of other complex systems, and often have many levels of organisation that exhibit what Herbert Simon (1962) called 'hierarchical organisation' (the hierarchy is of system and subsystem). There is an emergent hierarchy in the life sciences, because living systems and collectives are organised into different units that have different roles and that can be said to represent the state of the environment and internal states of the system itself. The ultimate result of all the features of a complex system above is an entity that is organised into a variety of levels of structure and properties; these interact with the level above and below and exhibit law-like and causal regularities and various kinds of symmetry, order and periodic behaviour. The best example of this is the whole system of life on Earth. Other systems that display such organisation include individual organisms, the brain, the cells of complex organisms and so on. 'Modularity' includes the differentiation of structure within a system, like the clustering of connections in a network, and also the organisation of the parts of a system to perform different functions, which often goes along with corresponding structural modularity (see Chapter 4).

3.10 History and Memory

Complex systems have order and structure that persist, and they also produce order and structure that persist. Persistence is always over some relevant time scale. It seems that nothing lasts forever. Eventually plants and animals die and their bodies break down. Even stars and galaxies have a finite, albeit very long, existence. The fate of the universe itself is unknown, but the things within it carry parts of its history with them because their complex structure is a product of that history and could not exist without it (as explained in Section 2.2). The Earth, with its heavy elements, is the product of and carries information about the history of the evolution of the universe over several generations of galaxies.

Living complex systems evolve over countless iterations of life cycles of generation and corruption. Genetic inheritance carries part of that history from parent to offspring. Every multicellular organism contains basic cellular mechanisms that were features of the most primitive life on Earth.

As discussed in Chapter 4, Bennett proposed logical depth to measure the history of complex systems. Information theory can readily be used to quantify history as correlations over time. "A system remembers through the persistence of internal structure" (Holland 1992). Any robust order that exists in a system can be thought of as memory, but in general we refer to it as 'history' and reserve the term 'memory' for something more specific. It is not helpful to think of a footprint in the sand as the beach's memory of a visitor because the footprint does not cause the beach to do anything differently. However, if an animal remembers the smell of a relative, this affects its behaviour. Hence, we use 'memory' only in relation to systems that exhibit adaptive behaviour and have internal degrees of freedom that are used to represent how the state of the environment or the system itself was at some time for the purposes of information processing.

3.11 Adaptive Behaviour

Adaptive behaviour is what organisms do to survive and reproduce. It ranges from the relatively simple and inflexible behaviour of a bacterium swimming up a chemical gradient in search of nutrients to the most sophisticated human deliberation. The American complexity scientist John Holland defines systems with adaptive behaviour as "systems [that] change and reorganize their component parts to adapt themselves to the problems posed by their surroundings." (Holland 1992).

Living complex systems organise themselves to do something. For example, neurons perform computational tasks, birds flock to migrate and army ants spontaneously build bridges over obstacles when foraging. So adaptive behaviour performs some function. Numerous kinds of adaptive behaviour are studied in biology, and many kinds of adaptive behaviour occur not just in living systems, but also in those complex systems derivative of life discussed in Chapter 2. Whether or not it is appropriate to attribute functions to physical systems that are not either biological or derivative of biological systems is highly contentious. We reserve the term 'adaptive behaviour' for living complex systems and complex systems produced by living complex systems. Collective motion is one of the main examples of adaptive behaviour in living systems, but recall that eusocial insects and other social organisms display many other forms of collectivity, such as the collective behaviour of bees cooling down their hive when it is too hot and the collective decision-making about the site of a new nest or whether or not to tie up ants in building a bridge over an obstacle rather than going around it. The brain can be modelled as neurons collectively making decisions.

The complex systems human beings have built all have a purpose or function. For example, the Internet is used to send information, and markets are for the exchange of goods and services between individuals or groups. The adaptive 'behaviour' which arises is communication efficiency, and the feedback between traders' actions and prices leading to markets finding stable prices. Adaptation may take the form of optimisation, but results are usually less than optimal. The collective functioning of all the nodes of the Internet or other networks, and even evolutionary change in natural or artificial systems, can all be modelled as emergent adaptive behaviour and decision making. All the products of complex systems except nonlinearity have functional versions of various kinds. Functional organisation is the functional version of order. Robustness often means retaining the structure that is required to perform some function or maintaining a parameter such as temperature. Robustness can also be functional robustness in the presence of structural change, as when the Internet is said to be robust because traffic is automatically rerouted when a part of the system fails. Modularity is the functional version of nested structure, as with, for example, the modularity of the Internet. As noted above, memory is best thought of as a functional notion. For example, the immune system remembers how to fight antigens it has encountered before because it has subsystems whose physical degrees of freedom can be copied to prepare antibodies. Most human artefacts are made to perform a function, and even art can be construed as providing for some kind of need. People readily think in functional terms, and many of the concepts of complexity science, such as information, computation, organisation and memory are functional. In fact, it is hard not to slip into functional ways of thinking because they are so important and so useful. Even the climate can be thought of as having a function since it provides and maintains an environment for life.

3.12 Different Views of Complexity

There is a long-standing debate about whether or not emergent entities and properties are real in the way that material things are. For example, are cash flows real? They might instead be regarded as just emergent levels of description that should not be taken as describing real things in the way that physics does. For example, if one plays a video game, one imagines that a player hitting a ball on the screen is what makes the ball move, but of course there is no causal interaction between the separate parts of the screen, only between the circuits in the chips of the computer generating the image. This may explain why in quote (2) (on Page 63) the idea of a subjective description is raised. Note, however, that most of physics is about emergent entities but few would want to say that, for example, the solar wind is not real. Chapter 5 returns to this issue.

Emergence is often contrasted with reduction, and, as noted in Chapter 1 people often associate complexity science with the failure of reduction. For example, Anderson (1972) talks about the limitations of reductive methods in describing systems composed of many parts. Indeed, many treatments of complex systems come close to defining complexity science in terms of the rejection of reductionism. For example, Melanie Mitchell (2011) begins her book *Complexity* with a discussion of antireductionism, which she defines as the view that "the whole is more than the sum of its parts." Yet in the very same paragraph she goes on to say how complexity science "explain how complex behaviour can arise from large collections of simpler components." This paradox runs throughout the literature on complex systems. On the one hand, there are calls for reductionism to be rejected, yet on the other hand, complexity science is all about breaking systems into parts and explaining the behaviour of the whole that results. The paradox is resolved by the fact that it is the interactions among the parts that make for the behaviour of the whole.

There are many different conceptions of complexity. Considering which of the examples of complex systems in Chapter 2 have which of the features above allows for several viable conceptions of complexity. The most inclusive conception of complexity is that it is just emergence, and it is exhibited in different ways by each of the examples. In fact, it is exhibited by all matter due to its emergent properties and hierarchical structure from the subatomic scale upwards. Even a monatomic gas at constant pressure, temperature and volume exhibits the emergent order of the ideal gas laws, even though there is no order at the level of the individual molecules. A more restrictive view that is still very broad is that complexity is self-organisation, where this is a special case of emergence. This view excludes isolated systems of matter that only exhibit emergence of the simplest kind but includes many systems in condensed matter and the universe, as well as all living systems.

Many of the products of complexity, such as order and organisation, memory and robustness, as well as some of the conditions, such as disorder and feedback, are readily described in computational/information-theoretic language. To understand complex systems it is often possible and indeed necessary to ignore the details of the physical interactions among the parts. All that seems to matter are the abstract properties of these interactions, not exactly how they work. So, for example, the swarming behaviour of insects can be understood without regard to whether the individual animals are signalling to each other by scent, sight or sound. The crucial thing is that they are able to exchange information about their behaviour and the state of the environment. Complex systems are often thought of as maintaining their order and producing order by the exchange of information among their parts. Abstracting systems to networks also facilitates describing them in computational terms, and (as discussed above) models of very different systems in terms of their parts and the interconnections between them exhibit kinds of invariance and forms of universal behaviour. There are many computation/information-based conceptions of complexity and complex systems (associated with some of the measures discussed in Chapter 4).

Numerosity, diversity and disorder, feedback, non-equilibrium and non-linearity are all features that can be had by nonliving physical systems without functions, goals or purposes. On the other hand, adaptive behaviour is shown only by biological systems or those derivative of them. As noted in Chapter 1, many conceptions of complexity confine it to systems that exhibit adaptive behaviour. For example, Mitchell's (2011) examples of complex systems are all living systems (insect colonies) or parts of livings systems (the immune system, the brain) or derivative of living systems (economies, the World Wide Web) because she takes complex systems to be restricted to those that adapt. We call this the 'functional' view of complexity, because the idea of adaptation is applicable only to biological systems or other systems that perform functions to achieve goals or purposes which are nonliving systems made and sustained by living systems. Chapter 4 argues that the various measures of complexity in the literature do not measure complexity as such but measure different forms of the various features of complex systems explained in this chapter. Chapter 5 returns to the big questions about complexity and which if any of the above views is correct.

Chapter 4

Measuring Features of Complex Systems

This chapter is a guide to quantifying complexity based on the fruits of the analysis of the previous chapters. Many measures of complexity have been proposed since scientists first began to study complex systems, and the list is still growing. The main lesson of Chapter 3 is that complexity is a multi-faceted phenomenon and that complex systems have a variety of features not all of which are found in all of them. This implies that assigning a single number to complexity cannot do it justice. As the physicist and Nobel laureate Murray Gell-Mann noted early on, "A variety of different measures would be required to capture all our intuitive ideas about what is meant by complexity" (1995, p. 1).

If complexity is a collection of features rather than a single phenomenon, then all quantitative measures of complexity can quantify only aspects of complexity rather than complexity as such. This insight makes it prudent to ask what any purported 'measure of complexity' actually measures. In the final section of this chapter, a few, by now classic, measures of complexity from the 1980s and 1990s, mentioned in many discussions on the subject, are discussed, including effective complexity, effective measure complexity, statistical complexity, and logical depth. The discussion confirms that they each quantify one or two of the features identified in Chapter 3.

The fact that complexity measures ever quantify only one or two features of complexity and never the phenomenon as a whole should inform any practitioner's approach to quantifying complexity. The chapter goes through the features of complex systems identified in Chapter 3 and discusses mathematical means available to quantify them, accompanied by examples. A feature can take different forms in different scientific disciplines, and, therefore, there often exist several ways by which to measure any given feature. The techniques were often invented in a different discipline from those in which they are now applied as the provided examples from the scientific literature illustrate.

This chapter demonstrates the truism of complexity science that it is computational and probabilistic (truism 7 in Chapter 1). It also further explains some of the new kinds of invariance and forms of universal behaviour that emerge when complex systems are modelled as networks and information-processing systems (truisms 5 and 6). The distinction between the order that complex systems produce and the complex systems themselves is central to the analysis (truism 9).

The 'classic' measures that are discussed in the final section of this chapter were constructed as thought experiments rather than as measures to be applied to real-world systems. Hence, they do not belong in a practitioner's tool kit. However, they have played a role in developing our understanding of complexity over the decades, and they serve now to help us understand the distinction between measuring complexity and measuring features of complexity.

This chapter is accessible to a general reader, with the possible exception of Section 4.10, which contains the discussion of the classic measures of complexity. It is a more technical section compared to the rest of this chapter. Throughout this chapter, terminology is explained when it is used for the first time. Further mathematical background is given in the Appendix.

4.1 Numerosity

The most basic measure of complexity science is the counting of entities and of interactions between them. Numerosity is the oldest quantity in the history of science, and counting is among the most basic scientific methods. Counting is the foundation of measurement because quantities of everyday relevance such as length, mass and time can be counted in units such as metres, grams and seconds. Counting alone does not tell us what counts as 'more' in the sense of 'more is different', because, as noted in Chapter 3, when we consider emergent behaviour, how many is 'enough' depends on the system. For some systems it is the high number of elements that is relevant for complexity, as in fluid dynamical systems; for others it is the high number of interactions, as in a small group of swarming animals or small insect colonies; or it is both, as in the brain and many (if not most) complex systems. The number of interactions is as important as the number of elements in the system.

4.2 Disorder and Diversity

Disorder and diversity are related, and the words used to describe them overlap and are often not clearly defined. 'Disorder' usually refers to randomness, which is to say lack of correlation or structure. Disorder is therefore just the lack of order. It is worth stating this explicitly since it follows that any measure of order can be turned into a measure of disorder and vice versa.

A disordered system is one that is random in the sense of lacking correlations between parts over space or time or both, at least to some extent. It is worth remembering that complex systems are never completely disordered. In complex systems, disorder can exist at the lower level in terms of the stochasticity in the interactions between the parts, as well as at the higher level, in terms of the structure which emerges from them and which is never perfect. Thermal fluctuations are a form of disorder relevant to the dynamics of complex systems. For example, as noted in Chapter 3, thermal fluctuations are necessary for most biochemical processes to take place. The term 'noise' or 'thermal noise' is used more frequently than 'disorder' in this context.

A real or purely mathematical random system would not be described as 'diverse'. Instead, the term 'diversity' is often used to describe inhomogeneity in element type – that is types of different kinds. Measures have been designed specifically to address diversity in this sense. Some of these are discussed at the end of this section.

Interactions can be disordered in time or in their nature. Elements can be disordered in terms of type. The structure formed by a complex system can be disordered in its spatial configuration. All these kinds of disorder are relevant, and all are quantifiable.

Mathematically, disorder is described with the language of probability (see Section A in the Appendix for a brief introduction to probability theory). The elements or interactions which are disordered are represented as a random variable X with probability distribution P over the set $\mathscr{X}$ of possible events x (events are elements or interactions). A standard measures of disorder is the *variance*. The variance can be used for events that are numeric, such as the number of edges per node in a network, but not for types, such as species in a population. The variance of a random variable X,

$$\operatorname{Var} X := \mathbb{E}[(X - \mathbb{E}[X])^2] , \tag{4.1}$$

measures the average deviation from the mean. The equivalent notation in the physics literature is $\operatorname{Var} X = \langle (X - \langle X \rangle)^2 \rangle$. The broader a distribution of possible event values is the higher, in general, the variance. A second standard measure of disorder, the *Shannon entropy*, is a function from information

89

theory (see Section B in the Appendix for a brief introduction to information theory). The Shannon entropy of a random variable X with probability distribution P over events x is defined as

$$H(X) := - \sum_{x \in \mathcal{X}} P(x) \log P(x) . \qquad (4.2)$$

The Shannon entropy measures the amount of uncertainty in the probability distribution P. In the case of all probabilities being equal, the distribution is a so-called uniform distribution. In this case all events are equally likely, and the uncertainty, and hence the Shannon entropy, over events is maximal. The Shannon entropy is zero when one probability is one and the others are zero. If, for example, the events x were the possible outcomes of an election, then $H(X)$ would quantify the difficulty in predicting the actual outcome.

To illustrate these measures of disorder consider a network, such as the World Wide Web or a neural network. The disorder relevant to a network is structural disorder. A network with many nodes and edges between every pair of nodes is considered a network with no disorder. The origin of a given network structure is often studied with network-formation models. For an overview of this and other network-formation models see, for example, (Newman 2010). One of the first network-formation models is the so-called Erdös-Renyi random graph model (or just random graph model, Poisson model, or Bernoulli random graph) (Erdös and Rényi 1960). The Erdös-Renyi model is parametrised by the number of nodes n, the maximum number of edges M, and a parameter p which is the probability of an edge being created between two existing nodes. Initially, the network has n nodes and no edges. In a subsequent formation process, with probability p, two nodes are connected by an edge. When $p = 0$, the resulting network after many repetitions is a set of nodes without any edges. For $p = 1$, the result is a highly connected network. For p somewhere in between 0 and 1, the formation process yields a network with links between some nodes and some nodes having more links than others. In this case, the probabilistic nature of the link formation results in a disordered structure of the network. Hence, the disorder of the formation process is taken as a proxy for the disorder of the final network structure. Several properties of the fully formed network, such as the average path length and the average number of edges per node, can be expressed as functions of n, M, and p only. These regularities emerge out of the disorder in the formation process.

The variance can be used to quantify the disorder of the network-formation process after assigning numeric values to the events 'edge' and 'no edge' – for example 1 and 0, respectively. The variance of the binary probability distribution $P = \{p, 1 - p\}$ of the Erdös-Renyi random graph model is

$\text{Var}\,X = p(1-p)$, which is maximal for $p = 1/2$. The Shannon entropy of the network-formation process can be computed without assigning numerical values to the events. The Shannon entropy of the binary probability distribution $P = \{p, 1-p\}$ of the Erdös-Renyi random graph model is $H(X) = -p\log p - (1-p)\log(1-p)$, which is also maximal for $p = 1/2$. Both measures are zero when $p = 0$ and, due to symmetry, when $p = 1$. If one were to measure the disorder in the final network structure itself, the variance and the Shannon entropy should be computed from the probability distribution over the node degrees. The result would be equivalent to the previous one in the sense that the degree distribution is trivial for $p = 0$ (completely disconnected) and $p = 1$ (fully or nearly fully connected), in which case both measures yield the value 0. For non-trivial network structures both measures are non-zero. It was remarked above that a measure of order can be used as a measure of lack of disorder and vice versa. Hence, any of the existing measures of network structure, such as average path length or clustering, can be used to measure disorder by monitoring their change. This approach to measuring disorder has been used in the study of Alzheimer's disease and its effect on neural connectivity in the human brain (see Bullmore and Sporns (2012) and references therein).

Temporal disorder in a sequence of events, such as the sequence of daily share prices on a stock market, is described with the language of stochastic processes. Disorder in a stochastic process is the lack of correlations between past and future events. A stochastic process is defined as a sequence of random variables X_t ordered in time (see Section A of the Appendix for more details). Disorder is the lack of predictability of future events when past events are known. To quantify disorder in a sequence $X_1 X_2 \ldots X_n$, the joint probability over two or more of the random variables is required, written as $P(X_1 X_2 \ldots X_n)$. This is the probability of the events occurring together (jointly). When a joint probability of two events is known, then, in addition to their individual probability, it is known how likely they are to occur together. An example is the probability of certain genetic mutations being present and the probability of two mutations being present in the same genome. The joint Shannon entropy $H(X_1 X_2 \ldots X_n)$ over this distribution,

$$H(X_1 X_2 \ldots X_n) := - \sum_{x^n \in \mathscr{X}^n} P(x_1 x_2 \ldots x_n) \log P(x_1 x_2 \ldots x_n) \,, \qquad (4.3)$$

captures the lack of correlations. A measure of average temporal disorder is the so-called *Shannon entropy rate*,

$$h_n : \frac{1}{n} H(X_1 X_2 \ldots X_n) \,. \qquad (4.4)$$

91

The Shannon entropy rate measures the uncertainty in the next event, X_n, given that all $n - 1$ previous events $X_1 \ldots X_{n-1}$ have been observed. The lower the entropy rate, the more correlations there are between past and future events and the more predictable the process is.

A fly's brain is an example of a complex system where temporal disorder has been measured experimentally. van Steveninck et al. (1997) recorded spike trains of a motion-sensitive neuron in the fly's visual system. From repeated recordings of neural spike trains, they constructed a probability distribution $P(X_1 X_2 \ldots X_k)$ over spike trains of some length k. From this probability distribution they computed the joint Shannon entropy $H(X_1 X_2 \ldots X_k)$ and the entropy rate $\frac{1}{k} H(X_1 X_2 \ldots X_k)$. They repeated the experiments after exposing the fly to the controlled external stimulus of a visual image and computed the Shannon entropy and entropy rate again. They interpreted the difference in the entropies between the two experiments as the reduction in disorder of the neural firing signal when a stimulus is present.

One speaks of the 'diversity' of species in an ecosystem or of diversity of stocks in an investment portfolio rather than 'disorder'. In the language of diversity, the elements, species or stocks, are called 'types'. The simplest measure of diversity is the number of types or the logarithm of that number. A more informative measure takes into account the frequency of each type, this being the number of individuals of each species in a habitat or the number of each stock in the portfolio. Treating such frequencies as probabilities, a random variable X of types $\mathscr{X}$ can be constructed, and the Shannon entropy $H(X)$ is used as a measure of type diversity. In ecology, diversity is measured using the *entropy power*, $2^{H(X)}$ (if the entropy is computed using log base 2 or $e^{H(X)}$ if the entropy is computed using log base e) (Jost 2006). It behaves similar to the entropy itself but has a useful interpretation: the entropy power is the number of species in a hypothetical population in which all species are equally abundant and whose species distribution has the same Shannon entropy as the actual distribution. If the types are numeric, such as the size of pups in an elephant seal colony (Fabiani et al. 2004), diversity can be measured using the variance. Often a normalised form of variance, the *coefficient of variation*, is used:

$$\mathrm{cv} := \frac{\sqrt{\mathrm{Var}X}}{\langle X \rangle}. \tag{4.5}$$

The coefficient of variation is the square root of the variance (also known as the standard deviation) divided by the mean. Its behaviour is equivalent to that of the variance. Broader distributions, such as a larger range of pup sizes in an elephant seal colony, result in a higher coefficient of variation.

However, it allows the comparison of distributions with the same variance but different means. A distribution with a variance of 10 and a mean of 20 might be considered more diverse than a distribution with a variance of 10 and a mean of 1,000. Their coefficient of variation would reflect this difference.

Scott Page, in his book 'Diversity and Complexity' (2010), distinguishes between three kinds of diversity: diversity within a type, diversity across types, and diversity of community composition. All three are measured by the Shannon entropy. In fact, they differ only in what constitutes an event in the definition of the random variable. Other measures of diversity are the so-called 'distance' measures and 'attribute' measures. Distance measures of diversity, such as the Weitzman Diversity, take into account not only the number of types, but also how much they differ from each other (Weitzman 1992) and therefor require a mathematical measure of distance. Attribute-diversity measures assign attributes to each type and numerically weigh the importance of each attribute. For example, to compute an attribute diversity of phenotypes more weight is put on traits with higher relevance for survival (see Page (2010) for more details).

4.3 Feedback

The interactions in complex systems are iterated so that there is feedback from previous interactions, in the sense that the parts of the system interact with their neighbours at later times depending on how they interacted with them at earlier times. And these interactions take place over a similar time scale to that of the dynamics of the system as a whole. There is no measure of feedback as such. Instead, the effects of feedback such as nonlinearity or structure formation are measured. Hence, the mathematical tools that are used to measure order and nonlinearity, as described later in the chapter (Sections 4.5 and 4.6), can also be indicators of feedback.

A common way to study feedback is to construct a mathematical model with feedback built into it. If the model reproduces the observed dynamics well, this suggests the presence of feedback in the system that is being modelled. An example is the dynamics of a population of predator and prey species such as foxes and rabbits. The growth and decline of these species can be modelled by the Lotka-Volterra differential equation model. It describes the change over time in population size of two species, the prey x and its predator y, using the four parameters A, B, C, and D. A and C are parameters for the speed of growth of the respective species. B and D quantify the predation. The change over time in population size, $\dot{x}$ and $\dot{y}$, is given by the two coupled equations

$$\dot{x} = Ax - Bxy \,,$$
$$\dot{y} = -Cy + Dxy \,. \tag{4.6}$$

The fact that x and y appear in both equations ensures that there is a feedback between the size of each population. If B or D are zero, there is no feedback.

For certain values of the parameters A, B, C, and D the number of individuals of each species oscillates. When the overabundance of predators reduces the number of prey to below the level needed to sustain the predator population but the resulting decline in the number of predators allows the prey to recover, a cycle of growth and decline results. For such oscillations to happen the time scale of growth, captured by A and C, needs to be similar to the time scale of predation, captured by B and D. Oscillations in predator-prey populations is a classic example of feedback.

A widely used computational tool for studying feedback are so-called *agent-based models*. These models are computational simulations of agents undergoing repeated interactions following simple rules. In such a simulation a usually large set of agents is equipped with a small set of actions that each agent is allowed to execute and a small set of (usually simple) rules defining the interaction between the agents. In any given round of a simulation an agent and an action, or two agents and an interaction, are picked at random. If the action (interaction) is allowed, it is executed. An agent-based simulation usually consists of many thousands of such rounds. One of the first agent-based models was the sugarscape model, pioneered by the American epidemiologist Joshua Epstein and computational, social and political scientist Robert Axtell (Epstein and Axtell 1996). The sugarscape model is a grid of cells, some of which contain 'sugar'; the others contain nothing. Agents 'move' on this landscape of cells and 'eat' when they find a cell containing sugar. Even this very simple setup produces emergent phenomena such as the feedback effect of the-rich-get-richer, which was described in Sections 2.5 and 2.6 of Chapter 2.

Agent-based models are frequently used to study feedback in the coherent dynamics of animal groups (Couzin and Krause 2002). Couzin and Franks (2003) describe observations of army ants in Soberania National Park in Panama. Army ants make an excellent study case for collective phenomena since they are able to form large-scale traffic lanes to transport food and building material over long distances. They even form bridges out of ants to avoid 'traffic congestion'. These collective phenomena are impossible without the presence of feedback. The authors set up an agent-based simulation with simple movement and interaction rules for individual ants. Feedback is built in as an ant's tendency to avoid collision with other ants and in its response to local pheromone concentration. The simulations reproduce the observed lane formations and the minimisation of congestion. Such a simulation is not to be confused with the measurement of actual feedback in a real system.

There are other notions of feedback in the literature on complex systems. The computational notion of feedback is to 'feed back' the output of a computation as input into the same computation. In this way, the outcome of future computations depends on the outcome of previous computations. This kind of feedback is particularly important for those who view nature to be inherently computational (Davies and Gregersen 2014; Lloyd 2006). On this view, any loop in the computational representation of a natural system indicates the presence of feedback. Nobel Laureate Paul Nurse made a similar point when presenting his computational view of the cell (Nurse 2008).

The above tools for analysing feedback have in common that they do not assign a number to the phenomenon, as is done in the case of disorder or diversity. Instead, in most practical applications feedback is a tunable interaction parameter of a model or an observable consequence of the interactions which are programmed into a model.

4.4 Non-Equilibrium

Complex systems are open to the environment, and they are often driven by something external. Non-equilibrium physical systems are treated by the theories of non-equilibrium thermodynamics (De Groot and Mazur 2013) and stochastic processes (Van Kampen 1992). Stochastic complex systems, such as chemical reaction systems, are often studied using the statistics of Markov chains. Consider a system represented by a set of states, S, through which the system evolves in discrete time steps. Let $\{P_{ij}\}$ be a matrix of time-independent probabilities of transitioning from state i to state j, with $\sum_j P_{ij} = 1$,[1] for all $i \in S$. Let π_i be the probability of being in state i. If there exists a probability distribution π^* such that, for all j,

$$\pi_j^* = \sum_i P_{ij} \pi_i^* , \qquad (4.7)$$

it is called the invariant distribution. In a stochastic model of a system of chemical reactions, for example, the chemical composition is represented as a probability distribution, and chemical reactions are represented as stochastic transitions from one reactant to another. A system is in chemical equilib-

---

[1] Some scientific fields use the reverse order, $\{P_{ji}\}$.

rium if the chemical composition is time-invariant. Reactions are still taking place in chemical equilibrium, but the depletion of one reactant is compensated by other transformations such that the overall concentrations remain largely unchanged. A general framework to model non-equilibrium stochastic dynamical systems is that of stochastic differential equations (Ikeda and Watanabe 2014).

4.5   Spontaneous Order and Self-Organisation

Perhaps the most fundamental idea in complexity science is that of order in a system's structure or behaviour that arises from the aggregate of a very large number of disordered and uncoordinated interactions between elements. Such *self-organisation* can be quantified by measuring the order that results – for example, the order in some data about the system. However, measures of order are not measures of self-organisation as such since they cannot determine how the order arose. This is because the order in a string of numbers is the same regardless of its source. Whether the order is produced spontaneously as a result of uncoordinated interactions in the system or whether it is the result of external control cannot be inferred from measuring the order without background knowledge about the system. For example, the orderly traffic lanes to and from food sources formed by an ant colony are considered the result of a self-organising process since there is no mechanism which centrally controls the ants' behaviour, while the orderly checkout lines in a supermarket are the result of a centrally managed control system. A high measure of order, even when self-organised, is not to be confused with a high level of complexity since order is but one aspect of complexity. However, the plethora of measures of order which are labelled as measures of complexity reflects the ubiquity of order in complex systems and explains the frequent use of order as a proxy for complexity.

Complex systems can produce order in their environment. It is important to remember that the order produced by the system is different from the order in the system itself. For example, the order of hexagonal wax cells built by honey bees is order produced by the system, while division of labour in the hive is order in the system. The hexagonal honeycomb structures are a form of spatial correlation which can be quantified by correlation measures, some of which are discussed in the following.

A correlation function is a means to measure dependence between random variables; therefore, it is a statistical measure. The *covariance* is a standard measure of correlation. For any two numeric random variables X and

Y, the covariance,

$$\mathrm{cov}(X,Y) = \mathbb{E}[XY] - \mathbb{E}[X]\mathbb{E}[Y] , \tag{4.8}$$

is the difference between the product of the expectations and the expectation of the product. If the two random variables are uncorrelated, this difference is zero. From the covariance a dimensionless correlation measure is derived, the so-called *Pearson correlation*. It is the most standard measure of correlation and defined as follows:

$$\mathrm{corr}(X,Y) := \frac{\mathrm{cov}(X,Y)}{\sigma_X \sigma_Y} , \tag{4.9}$$

where σ is the square root of the variance, known as the standard deviation.

A measure of correlation derived from information theory is the *mutual information*. For two random variables X and Y, the mutual information is a function of the Shannon entropy H (see Section 4.2):

$$I(X;Y) = H(X) + H(Y) - H(X,Y) . \tag{4.10}$$

The mutual information measures the difference in uncertainty between the sum of the individual random variable distributions and the joint distribution. If there are any correlations between the two variables, the uncertainty in their joint distribution will be lower than the sum of the individual distributions. This is a mathematical version of the often repeated statement that 'the whole is more than the sum of its parts'. If the whole is different from the sum of the parts, it means that there are correlations between the parts. For two completely independent random variables, on the other hand, $H(X) + H(Y) = H(X,Y)$ and the mutual information is zero.

An example of covariance as a measure of order is the study of bird flocking by William Bialek, Andrea Cavagna, and colleagues (2012). They filmed flocks of starlings in the sky of Rome (containing thousands of starlings) and extracted the flight paths of the individual birds from these videos. Each bird's different flight directions over time were represented as a random variable, and the random variables of all birds were used to compute their pairwise covariances.[2] This list of covariances was fed into a computer simulation that modelled the flock of birds as a condensed matter system, which is defined by the interaction between close-by 'atoms' only. The computer simulation of such a very simple model with pairwise interactions only and no

---

[2]They used the convention from statistical mechanics in which the uncorrelated average product is not subtracted. Thus, their covariance is the statistical mechanical correlation function $\mathbb{E}[XY]$.

further parameters, produced a self-organising system that very closely resembled the self-organising movement originally observed. A similar analysis was been done on network data of cultured cortical neurons, corroborating the idea that the brain is self-organising (Schneidman et al. 2006).

The order in a flock of starlings is a spatial order persistent over time. Systems in which the focus is more on the temporal aspect of the order are neurons and their spiking sequences, for example, or the monsoon season and its patterns. Order in these systems is studied by representing them as sequences of random variables $X_1 X_2 \ldots X_t$ with a joint probability distribution $P(X_1 X_2 \ldots X_t)$. Such sequences we encountered above in the study of disorder. Several authors, independently, introduced the mutual information between parts of a sequence of random variables as a measure of order in complex systems, under the names of *effective measure complexity* (EMC) (Grassberger 1986), *predictive information* (I_{pred}) (Bialek et al. 2001), and *excess entropy* (E) (Crutchfield and Feldman 2003). Consider the infinite sequence of random variables $X_{-t} X_{-t+1} \ldots X_0 X_1 X_2 \ldots X_t$, which is also called a stochastic process. The information theoretic measure I_{pred} (or EMC or E) of correlation between the two halves of a stochastic process is defined as the mutual information between the two halves:

$$I_{pred} := \lim_{t \to \infty} I(X_{-t} X_{-t+1} \ldots X_{-1} ; X_0 X_1 \ldots X_t) . \tag{4.11}$$

There is, of course, never an infinite time course of data, and the limit $t \to \infty$ is never taken in practice.

Palmer et al. (2015) measured the predictive information in retinal ganglion cells of salamanders. Ganglion cells are a type of neuron located near the inner surface of the retina of the eye. In the lab, the salamanders were exposed, alternatively, to videos of natural scenes and to a video of random flickering. While a video was showing, the researchers recorded a salamander's neural firings. Repeated experiments allowed them to infer the joint probability distribution $P(X_{-t} \ldots X_t)$ over the ganglion cell firing rates and to compute the predictive information I_{pred} contained in it. They found that I_{pred} was highest when a salamander was exposed to naturalistic videos of underwater scenes. This shows that the order in the natural scenes is reflected in the order of the neural spike sequences. The authors also think that it shows the neural system not only responds to a visual stimulus, but also makes predictions about it.

Quantifying predictability and actually predicting what a system is going to do are, of course, two different things. In order to make a prediction one first has to have a model, for example, inferred from a set of measured data.

4.6 Nonlinearity

There are several different phenomena addressed with the same label of 'non-linearity'. Each phenomenon requires its own measure. Power laws are probably the most prominent examples of nonlinearity in complex systems. But correlations as a form of nonlinearity are equally important, and these two are not completely separate phenomena.

4.6.1 Nonlinearity as Power Laws

A power law is a relation between two variables, x and y, such that y is a function of the power of x – for example, $y = x^\mu$. Quite a few phenomena in complex systems, such as the relation between metabolism and body mass or the number of taxpayers with a certain income and the amount of this income, follow a power law to some extent. The nonlinear relation between taxpayer bracket and number of people in this bracket was described in Chapter 2 in the context of economics. The power law of metabolism for mammals was first discussed by Max Kleiber (1932) in 1932. It is now well established that, to a surprising accuracy, the metabolic rate of mammals, R, is proportional to their body mass, m, to the power of $3/4$: $R \propto m^{3/4}$ (see West et al. (1997) and references therein). Because $3/4$ is less than 1, a mammal's metabolism is more efficient the bigger the mammal; an elephant requires less energy per unit mass than a mouse. This is a nonlinear effect since doubling the body size does not double the energy requirements. It is also another instance of the often repeated, but confused, statement that, in complex systems, the whole is more than the sum of its parts. The whole is never more than the sum of its parts when interactions are taken into account.

The relation between taxpayer bracket and number of people in this bracket is an instance of a statistical distribution that exhibits a power-law behaviour. Other examples of statistical distributions with a power-law behaviour are the number of metropolitan areas relative to their population size, the number of websites relative to the number of other websites linking to them, and the number of proteins relative to the number of other proteins that they interact with (for reviews, see Newman 2005; Sornette 2009).

Statistical distributions with a power-law behaviour are defined in terms of random variables. Consider a discrete random variable X with positive events $x > 0$ and probability distribution P. The distribution P follows a power law if

$$P(x) = cx^{-\gamma},\tag{4.12}$$

for some constant $\gamma > 1$ and normalisation constant $c = (\gamma - 1)/(x_{\min}^{1-\gamma})$, where $x_{\min}$ is the smallest of the x values. A cumulative distribution with a power-law form is also called a Pareto distribution (see Section 2.5); a discrete distribution with a power-law form is also called a Zipf distribution (for a review, see Mitzenmacher 2004). Eq. 4.12 can be written as $\log P(x) = \log c - \gamma \log x$, which says that plotting $\log P(x)$ versus $\log x$ yields a straight line with slope $-\gamma$. Therefore, the presence of a power law in real-world distributions is often determined by fitting a straight line to a log-log plot of the data. Although this is common practice, there are many problems with this method of identifying a power-law distribution (Clauset et al. 2009).

A power-law distribution has a well-defined mean for $\gamma \leq 1$ over $x \in [1, \infty)$ and a well-defined variance for $\gamma \leq 2$. Power-law distributions are members of the larger family of so-called fat-tailed distributions. These probability distributions are distinct from the most common distributions, such as the Gaussian or normal distribution, in that events far away from the mean have non-negligible probability. Such rare events have obtained the name 'black swan' events; they come as a surprise but have major consequences (Taleb 2007).

4.6.2 Nonlinearity versus Chaos

Nonlinearity in complex systems is not to be confused with nonlinearity in dynamical systems. Nonlinear dynamical systems are sets of equations, often deterministic, describing a trajectory in phase space, either continuous or discrete in time. Some of these systems exhibit chaos, which is the mathematical phenomenon of extreme sensitivity of the trajectory on initial conditions. An example of a discrete dynamical system exhibiting chaos is the logistic map, already encountered in Chapter 1. The logistic map, $x_{t+1} = rx_t(1-x_t)$ where t indexes time, is a simple model of population dynamics of a single species, as opposed to two species, discussed above in the context of feedback. This map is now a canonical example of chaos.

Actual physical systems studied by dynamical systems theory, such as a chaotic pendulum, need not have any of the features of complex systems. Certainly, chaos and complexity are two distinct phenomena. On the other hand, the time evolution of many complex systems is described by nonlinear equations. Some climate dynamics, for example, are modelled using the deterministic Navier-Stokes equations, which are a set of nonlinear equations describing the motion of fluids. Another example of a nonlinear equation used to describe many complex systems is the Fisher-KPP differential equation (Fisher 1937). Originally introduced in the context of population dynamics, its application ranges from plasma physics to physiology and ecology.

4.6.3 Nonlinearity as Correlations or Feedback

For some the notion of nonlinearity in complex systems is synonymous with the presence of correlations (for instance, MacKay 2008). If two random variables X and Y are independent, their joint probability distribution $P(XY)$ is equal to the product distribution $P(X)P(Y)$. When this equality does not hold, then there must be correlations between X and Y.

Defining 'nonlinearity' in terms of the presence of correlations is not to be confused with linear versus nonlinear correlations. In the language of statistical science, two variables X and Y are linearly correlated if one can be expressed as a scaled version of the other, $X = a + cY$, for some constants a and c. The Pearson correlation coefficient, for example, detects linear correlations only. The mutual information, on the other hand, detects all correlations, linear as well as nonlinear.

To others, mainly social scientists, 'nonlinearity' means that the causal links of the system form something more complicated than a single chain. A system with causal loops, indicating feedback, would count as 'nonlinear' in this view (Blalock 1985).

The different definitions of nonlinearity discussed here are all ubiquitous in complex systems research, so it is not surprising that nonlinearity is often mentioned as essential to complex systems.

4.7 Robustness

Several phenomena are often grouped together under the umbrella of 'robustness'. A system might be robust against perturbation in the sense of maintaining its structure or its function upon perturbation, which some refer to as 'stability'. Alternatively, a system might be robust in the sense that it is able to recover from a perturbation; this is also called 'resilience'.

Strictly speaking, robustness is the property of a model, an algorithm, or an experiment that is robust against the change of parameters, of input, or of assumptions. But usually, in the context of complex systems, robustness refers to the stability of structure, dynamics or behaviour in the presence of perturbation. All order and organisation must be robust to some extent to be worth studying. Several tools are available for studying robustness; the most frequently used are tools from dynamical systems theory and from the theory of phase transitions. As with all topics in this chapter, brief descriptions are given outlining the role of these tools in the study of complex systems.

4.7.1 Stability Analysis

The system of predator and prey species sharing a habitat, which was discussed above (see the Lotka-Volterra population model in Section 4.3 and
Section 4.6), is an example of a stable dynamical system. After some time
the proportion of the two species becomes either constant or oscillates regularly, independent of the exact proportion of species in the beginning. A
dynamical system is called 'stable' if it reaches the same equilibrium state
under different initial conditions or if it returns to the same equilibrium state
after a small perturbation. Stability analysis is prevalent in physics, nonlinear dynamics, chemistry, and ecology. A reversible chemical reaction, for
example, might be stable with respect to forced decrease or increase of a reactant, which means the proportion of reactants and products returns to the
same value as before the perturbation. Other examples of complex systems
which are represented as dynamical systems are food webs with more than
two species (Pimm 1982; Rooney et al. 2006), genetic regulatory networks
(de Jong 2002), and neural brain regions (Friston 2009).

For any given dynamical system described by a state vector $\mathbf{x}$ and a set of
(possibly coupled) differential equations $dx_i/dt = f_i(\mathbf{x})$, a stable point, a so-
called fixed point, is a solution to the equations $dx_i/dt = 0$. Stability analysis
classifies these fixed points into stable and unstable ones (or possibly stable
in one direction and unstable in another). Assuming the system is at one of
its fixed points, the effect of a small perturbation on the system's dynamics
is found by analysing the Jacobian matrix J, a linearisation of the system,
which is defined as

$$[J_{ij}] = \left[\frac{\partial f_i}{\partial x_j} \right]. \tag{4.13}$$

If the eigenvalues of the Jacobian evaluated at a given fixed point all have
real parts that are negative, then this point is a stable fixed point and the system returns to the steady state upon perturbation. If any eigenvalue has a real
part that is positive, then the fixed point is unstable and the system will move
away from the fixed point in the direction of the corresponding eigenvector,
usually towards another, stable, fixed point. For an introduction to fixed-point
analysis of dynamical systems, see, for example, Strogatz (2014). Any stable fixed point is embedded in a so-called basin of attraction. The size of this
basin quantifies the strength of the perturbation which the system can with-
stand and, therefore, is a measure of the stability of the system at the fixed
point (Strogatz 2014). Stability analysis is widely used in ecology (Holling
1973; Scheffer 2010).

Viability theory combines stability analysis of deterministic dynamical systems theory with control theory (Aubin 1990, 2009). It extends stability analysis to more general, non-deterministic systems and provides a mathematical framework for predicting the effect of controlled actions on the dynamics of such systems, with the aim of regulating them. Viability theory has been applied to the resilience of social-ecological systems (Béné and Doyen 2018; Deffuant and Gilbert 2011).

A similar, though mostly qualitative, use of the ideas of stability and viability is found in the analysis of *tipping points* in climate and ecosystems. Tipping points are the points of transition from one stable basin of attraction to another, instigated by external perturbations (Scheffer 2010).

4.7.2 Critical Slowing Down and Tipping Points

The time it takes for a system to return to a steady state after a perturbation is a stability indicator complementary to the fixed-point classification and the size of the attractor basin. The longer it takes the system to recover after a perturbation the more fragile the system is. An increase in relaxation time can indicate a *critical slowing down* and the vicinity to a so-called tipping point or phase transition. When a system is close to a tipping point, it does not recover anymore from even very small perturbations and moves to a different steady state which is possibly very far away from its previous state. Finding measurable indicators for nearby tipping points has been of considerable interest, in particular since ecological and climate systems have begun to be characterised by stability analysis and their fragility is being recognised more and more (Scheffer 2010; Scheffer et al. 2015).

Mathematically, the vicinity to a tipping point is recognised by the functional dependency of the recovery time on the perturbation strength. A system which is close to a tipping point exhibits a recovery time that grows proportional to perturbation strength to some power. This *scaling law*, associated with critical slowing down, is a well-known phenomenon in the statistical mechanics of phase transitions (see Section 2.1 of Chapter 2 for more on phase transitions). The standard example of a phase transition in physics is the magnetisation of a material as a function of temperature. The magnetisation density m is proportional to the power of the temperature difference to a critical temperature, $m \propto |T - T_C|^{-\alpha}$. T_C is the critical temperature, the equivalent to a tipping point, at which the magnetisation diverges and the system undergoes a phase transition.

Another signature of a nearby tipping point is an increase in fluctuations. In general, a perturbed system fluctuates around a steady state before settling back down. The larger the length or time scale on which the fluctuations are correlated, the closer the system is to a tipping point.

Experimentally, one might expose a system to increasingly strong perturbations and measure the time it takes the system to come back to its steady state. Such measurements yield the response of the system as a random variable S as a function of spatial coordinate $\mathbf{x}$ and time t. The covariance $\text{cov}(S(\mathbf{x},t), S(\mathbf{x}+\mathbf{r},t+\tau))$ (see Section 4.5) between the random variable at time t and spatial location $\mathbf{x}$ and the same variable at some later time $t+\tau$ and some displaced location $\mathbf{x}+\mathbf{r}$ is a measure of the temporal and spatial correlations in time. The equivalent measure in physics is the so-called *auto-correlation function*, denoted by $C(\mathbf{r}, \tau)$, defined, in physics notation, as

$$C(\mathbf{r}, \tau) = \langle S(\mathbf{x},t)S(\mathbf{x}+\mathbf{r},t+\tau)\rangle . \tag{4.14}$$

It differs from the covariance by not subtracting the product of the marginal expectations, $\langle S(\mathbf{x},t)\rangle\langle S(\mathbf{x}+\mathbf{r},t+\tau)\rangle$ (in statistics notation, $\mathbb{E}[S(\mathbf{x},t)]\mathbb{E}[S(\mathbf{x}+\mathbf{r},t+\tau)])$, from the expectation of the product (cf. eq. 4.8).

When correlations decay exponentially in time, C is proportional to $e^{-k\tau}$, where k is an inverse time. After time $\tau = 1/k$ correlations have decayed to a fraction $1/e$ of the value they had at time t, and $\tau = 1/k$ is the so-called characteristic time scale. Equally, when correlations decay exponentially with distance, the distance $|\mathbf{r}|$ at which they have decayed to a fraction $1/e$ of the value at $|\mathbf{r}| = 0$ is the characteristic length scale. Critical slowing down is accompanied by fluctuations that decay slower than exponentially. The signature in the auto-correlation function C is a power-law decay either in time or in space, $C \propto |\mathbf{r}|^{-\alpha}$ or $C \propto \tau^{-\alpha}$. Theoretically, at the point of a phase transition the correlation length becomes infinite. At that point the system has correlations on all scales and no characteristic length nor time scale anymore. A correlation length which captures nonlinear correlations has been based on the mutual information (Dunleavy et al. 2015).

An example of a complex system where critical slowing down has been measured is a population of cyanobacteria under increasing irradiation. The bacteria require light for photosynthesis, but irradiation levels that are too high are lethal. For protection against destructively high irradiation levels, bacteria have evolved a shielding mechanism. Annelies Veraart and colleagues exposed cell cultures of cyanobacteria to varying intensities of irradiation and studied the subsequent shielding process (2012). When the irradiation was relatively weak the bacterial population quickly recovered after enacting the mutual shielding mechanism by which the bacteria protect each other. The stronger the radiation, the longer it took the population to build up the necessary shielding and recover afterwards. Veraart and her colleagues measured a critical slowing down with a power-law-like behaviour. Once the light stress reached a certain threshold, equivalent to a critical point, the population collapsed. The new steady state that the population had tipped into was that of death. There are many other complex systems where critical slowing down has been suspected or observed – for example, in the food web of a lake after introduction of a predator species (Carpenter et al. 2011), in marine ecosystems in the Mediterranean after experimental removal of the algal canopy (Benedetti-Cecchi et al. 2015), and in paleoclimate data around the time of abrupt climatic shifts (Dakos et al. 2008). For a review of critical slowing down in ecosystems, see Scheffer et al. (2015). For many more examples of criticality in complex systems, ranging from geological to financial systems, see Sornette (2009).

4.7.3 Self-Organised Criticality and Scale Invariance

Power laws are an example of nonlinearity, as discussed in Section 4.6. Power-law behaviour is also an example of instability since a power-law behaviour in the recovery time is the signature of a system being driven towards a critical point, as discussed. It is, therefore, unexpected that many complex systems exhibit a power-law behaviour without any visible driving force and that they are nevertheless relatively stable. It appears that such systems stay close to a critical point 'by their own choice', a phenomenon called *self-organised criticality*. When it was discovered in a computer model of avalanches (Bak et al. (1988); see Section 2.1), it sparked a whole surge of studies into the mechanism behind self-organised criticality. This surge was fueled by experimental observations of power-law-like behaviour in a range of different systems, such as the Earth's mantle and the magnitude and timing of earthquakes and their afterquakes, or the brain and the timing of neurons (Bullmore and Sporns 2009; Sornette 2002; Zöller et al. 2001). In these systems, the relevant observable, magnitude or timing, was measured as a histogram of frequencies of events. The probability distribution $P(x)$ of events x, constructed from the data, decays approximately as a power law, $P(x) = cx^{-\gamma}$. As remarked above, the true functional form of these decays is still debated; it is rarely more than an approximate power law (Clauset et al. 2009). A power law implies so-called *scale invariance*, since ratios are invariant to scaling of the argument: $P(cx_1)/P(cx_2) = P(x_1)/P(x_2)$. Scale invariance has been observed in many natural as well as social complex systems (Sornette 2009), including scale invariance of the statistics of population sizes of cities (Bettencourt 2013).

While a power law in the auto-correlation function indicates instability and the vicinity of a critical point, a power law in a statistical distribution may indicate self-organised criticality which is associated with stability. Three decades after the discovery of self-organised criticality, there still is no known mechanism for it. The seeming contradiction between the robustness of a complex system, one of its emerging features, and the inherent instability of systems close to a tipping point remains unresolved. For a review of self-organised criticality, see Pruessner (2012) and Watkins et al. (2016).

4.7.4 Robustness of Complex Networks

Network structures are ubiquitous in the interactions within a complex system. It is therefore not surprising that complex networks have grown into their own subfield of complex systems research. Many examples of networks have been mentioned in this book, from protein-protein interactions and neural networks to financial networks and online social networks. A network is a collection of nodes connected via edges. The *degree* of a node is the number of edges connected to it. The nature of nodes and edges differs for each system. In protein-protein networks the nodes are proteins; two nodes are connected by an edge if they interact, either biochemically or through electrostatic forces. A *path* is a sequence of nodes such that every two consecutive nodes in the sequence are connected by an edge. The *path length* is the number of edges traversed along the sequence of a path. The average shortest path is the sum of all shortest path lengths divided by their number. The phrase 'six degrees of separation' refers to the average path length between nodes in social networks. This goes back to a now famous experiment performed by Stanley Milgram and his team in the 1960s (Milgram 1967). Milgram gave letters to participants randomly chosen from the population of the United States. The letters were addressed to a person unknown to them, and they were tasked with handing their letter to a person they knew by first name and who they believed would be more likely to know the recipient. This led to a chain of passings-on for each letter. Surprisingly, letters reached the addressee, on average, after only five intermediaries. The stability of average path length is one proxy for the robustness of a network. When edges or nodes are removed from the network and the average path length stays more or less the same, the network is considered robust (in this respect). Reka Albert, Hawoong Jeong and Albert-Lásló Barabási (2000) found that the Internet and the World Wide Web are very robust in precisely this way. The shortest path is hardly affected upon the random removal of nodes. Albert and her colleagues studied the structure of the World Wide Web and the Internet by taking real-world data and artificially removing nodes in a computer simulation. Plotting the shortest path against the fraction of nodes removed from the network revealed that the path length initially stayed approximately constant. Only once a large fraction of the nodes had been removed did the length suddenly and dramatically increase. This sudden increase is a form of phase transition between a well-connected phase and a disconnected phase. It is seen already in the simplest model of networks, the Erdös-Renyi random graph model discussed above (see Newman 2010). Other real-world networks exhibiting this structural form of robustness are protein networks (Jeong et al. 2001), food webs (Dunne et al. 2002), and social networks (Newman et al. 2002). Robustness is always with respect to a feature or function. Robustness with respect to one feature might not imply robustness with respect to another. The Internet, for example, is robust against random removal of nodes (servers), but it is considerably less robust to targeted removal of the highest-degree nodes.

4.8   Nested Structure and Modularity

Nested structure and modularity are two distinct phenomena, but they may be related. 'Nested structure' refers to structure on multiple scales. 'Modularity' is a functional division of labour, or specialisation of function among parts, or a structural modularity and frequently all of these together.

Structural modularity is a property much discussed especially in the context of networks, where it is referred to as 'clustering'. A cluster in a network is a collection of nodes that have many edges between one another compared to only few edges to nodes in the rest of the network. A simple example is the network of online social connections such as the network of 'friends' on Facebook. This network of social connections tends to be highly clustered since two 'friends' of any given user are more likely to also be 'friends' than to be unrelated.

Finding clusters in networks has received considerable attention, and many so-called clustering algorithms have been proposed. For an introduction to clustering algorithms, see, for example, Newman (2010). All clustering algorithms follow a similar principle. Given a network, they initially group the nodes into arbitrary communities, and, by some measure unique to each technique, they quantify the linking strength within each community and that in between communities. Information theoretic distance is one such measure (Rosvall and Bergstrom 2008). The algorithms then optimise the communities by moving nodes between them until the linking strength within each community is maximised and the linking strength in between communities is minimised. There is usually no unique solution to this optimisation problem, and the identified clusters might differ from algorithm to algorithm. The presence of clusters alone is not sufficient for modularity since the network could consist of one gigantic cluster, with every node being connected to most other nodes, and have no modularity at all.

Once a community structure of a network has been identified, the extent to which it is modular can then be quantified. One of the first measures designed to quantify structural modularity is the *modularity* measure by Mark Newman and Michelle Girvan (2004). It assumes that a community structure of a given network has been identified and that k clusters of nodes have been found. From these k clusters, a $k \times k$ matrix $\mathbf{e}$ is constructed in which the entries e_{ij} are the fraction of edges that link nodes in cluster i to nodes in cluster j. The matrix entries can be interpreted as the joint probabilities $\Pr(i, j)$ for the event of an edge to be attached to a node in cluster i and the joint event of this edge to end on a node in cluster j. If these two events are independent, the joint probability distribution is equal to its product distribution, $\Pr(i, j) = \Pr(i) \cdot \Pr(j)$. If, on the other hand, $\Pr(i, j) \neq \Pr(i) \cdot \Pr(j)$, then the probability $\Pr(i, j)$ is dependent on whether i and j are the same cluster ($i = j$) or not. With such a dependence present, there is modularity in the network. This condition of a joint probability distribution being a non-product distribution was a condition for 'nonlinearity as correlations' (Section 4.6.3).

Newman and Girvan use this dependency condition to define modularity Q as any deviation of the joint probability distribution $\Pr(i, i)$ of edges connecting nodes within the same cluster from the product distribution $\Pr(i) \cdot \Pr(i)$. In this sense, modularity is a form of nonlinearity as correlations. In the above matrix notation, the probability $\Pr(i) = \sum_j e_{ij}$. This can be understood as the so-called marginal probability of picking any edge in the network and for that edge to start in cluster i. Modularity is then defined as:

$$
Q := \sum_i \left(e_{ii} - \left(\sum_j e_{ij} \right)^2 \right) . \tag{4.15}
$$

This measure of modularity is also taken as an optimisation function for community detection algorithms, but limitations to its effectiveness have been pointed out (Brandes et al. 2007; Fortunato and Barthelemy 2007).

Many natural systems exhibit structure that is repeated again on a smaller scale; the structure is nested within itself. A cauliflower exhibits this particular form of spatial scale invariance in the structure of the florets consisting of smaller florets and so forth. Benoît Mandelbrot discovered the mathematics of such nested structures, for which he coined the term *fractal*. Fractals are mathematical objects with a perfect scale invariance, a repetition of structure at an infinite number of scales (Falconer 2004). Mandelbrot's now famous book, *The Fractal Geometry of Nature* (1983), revealed the ubiquitous presence of fractal structure in natural systems, both living and nonliving. Fractals have the mathematical property of a non-integer dimension, and therefore fractal dimension is sometimes used as an indicator of nested structure (e.g., in ecology; Sugihara and May 1990). For example, a circle has dimension 2, a sphere has dimension 3, and the dimension of a cauliflower is estimated at 2.8 (Kim 2004).

Another indicator for multiple scales is the power-law decay in a correlation function (see Section 4.7). For example, as mentioned in Section 2.6 of Chapter 2, the number of websites in the visible World Wide Web as a function of their degree approximately follows a power law with an exponent γ, which, in 1999, was estimated at 2.1 (Barabási and Albert 1999). This power-law decay is due to clusters of websites being nested within bigger clusters of websites. The World Wide Web has tens of billions of web pages, but only a few dozen domains own most of the links.[3] These central domains are linked to each other, as well as to web pages within their own domain, and they also connect to large clusters of less-well-connected domains. Each of these clusters has, again, a few highly connected domains. This structure of clusters of sites with a few highly linked domains repeats at ever smaller scales. This self-similar nesting of clusters is much studied in complex networks (Newman 2003; Ravasz and Barabási 2003). Methods based on statistical inference for identifying nested clusters have also been developed (Clauset et al. 2008).

The presence of scale invariance in the degree distribution of a network can be reproduced by a model of network growth first considered by Derek Price (1976). Starting with a small network, new nodes are added and connected by an edge to an existing node with a probability proportional to the existing node's degree. Hence, any new edge will affect the probability of future edges being added. Connecting a large number of nodes following this rule results in a network where a few nodes have a very high number of edges, and most nodes have very few. The algorithm is called the *preferential attachment algorithm*. It is a variant of a random graph model (see Section 4.2), and it describes the rich-get-richer effect seen in economics (see Section 2.5 of Chapter 2). It clearly has feedback built into it. The initial degree distribution might be uniformly random, but, after many iterations, it gets locked into a very skewed distribution due to the feedback of previously formed edges on future edge formation. The preferential attachment mechanism illustrates why power laws have been, and still are, a central theme in many studies of complex systems. Power-law-like behaviour, already discussed repeatedly in this chapter, can serve as an indicator for several of the features of complex systems identified in this book: nonlinearity, (lack of)

---

[3] A domain is the name you need to buy or register, such as google.com. Any web page with a url within this domain name, such as www.google.com/maps is part of this domain.

robustness, nested structure, and feedback. This also suggests that these phenomena are not isolated from each other.

4.9   History and Memory

The various measures of complexity measure different features of complex systems, all of which arise because of their histories. Hence, other measures can be used as proxies for history. For example, a network may have a definite growth rate so that the size of the network can be used as a measure of its age. Another way to measure the history of a complex system is to measure the structure it has left behind, because the more of it there is, the longer the history required for it to spontaneously arise as a result of the complex system's internal dynamics and interaction with the environment. However, in some cases, the structure in the world is built very quickly and deliberately rather than arising spontaneously, like a beaver's dam or a ploughed field. Background knowledge is needed to know how to relate such structure to history. There are no direct measures of history used in practice, but the logical depth discussed in the next section was introduced to capture the idea that complex systems require a long history to develop. Any measure of correlations in time, including the statistical complexity discussed below, can be considered a measure of memory.

4.10   Computational Measures

Many of the growing number of measures of complexity are based on computational concepts such as algorithmic complexity and compressibility.[4] The previous section showed that complexity measures capture features of complexity but not complexity as such. This section discusses measures that consider complex systems to be computational devices with memory and computational power. All of these measures are reminiscent of thought experiments in that they are not implementable in practice or even in principle. Although these measures are now decades old (and none measure complexity as such) they are included here because they have had a considerable influence on thinking about complexity. We explain what feature of complexity each measures.

---

[4]Lloyd (2001) produced a 'non-exhaustive list' of over forty, and many more measures have been defined since then.

4.10.1 Thermodynamic Depth

Thermodynamic depth was introduced by physicists Seth Lloyd and Heinz Pagels (1988). Lloyd and Pagels started out with the intuition that a complex system is neither perfectly ordered nor perfectly random and that a complex system plus a copy of it is not much more complex than one system alone. To specify the order of a complex system they consider the physical state of the system at time t_n, calling it s_n. In any stochastic setting, a given state can be preceded by more than one state. In other words, the set of states a system was in at times t_1 to t_{n-1}, a trajectory of length $n-1$, is not unique. Assigning a probability to any such trajectory which leads to state s_n, $\Pr(s_1, s_2, \ldots, s_{n-1}|s_n)$, the thermodynamic depth of state s_n is defined as $-k \ln \Pr(s_1, s_2, \ldots, s_{n-1}|s_n)$ averaged over all possible trajectories $s_1, s_2, \ldots, s_{n-1}$,

$$\mathscr{D}(s_n) = -k \sum_{s_1, \ldots, s_{n-1}} \Pr(s_1, s_2, \ldots, s_{n-1}|s_n) \ln \Pr(s_1, s_2, \ldots, s_{n-1}|s_n) , \quad (4.16)$$

where k is the Boltzmann constant from statistical mechanics. In this view, the complexity of a system is given by the thermodynamic depth of its state. The intuition that the thermodynamic depth is intended to capture is that systems with many possible and long histories are more complex than systems which have short, and thus necessarily fewer possible, histories. What this definition leaves open, and arguably subjective, is how to find the possible histories, their lengths and what probabilities to assign to them (Ladyman et al. 2013). Thus, practically, the measure is not implementable. The rate of increase of thermodynamic depth when considering histories further and further back in time is mathematically an entropy rate, which is a measure of disorder (see Section 4.2). Thus, while the intention was for thermodynamic depth to be a measure of history, it is in fact a measure of disorder. This was pointed out in Crutchfield and Shalizi (1999).

4.10.2 Statistical Complexity and True Measure Complexity

The *quantitative theory of self-generated complexity*, introduced by physicist Peter Grassberger (1986), and *computational mechanics*, introduced by physicists James Crutchfield and Karl Young (Crutchfield 1994; Crutchfield and Young 1989), are similar frameworks that go beyond providing a measure to inferring a computational representation for a complex system. The former comes with a measure called *true measure complexity*, the latter with a measure called *statistical complexity*. Since computational mechanics has been developed in more statistical and practical detail, we focus on it here (see Shalizi and Crutchfield 2001).

The assumption of computational mechanics is that a complex system is an information-storing and -processing entity. Hence, any structured behaviour it exhibits is the result of a computation. The starting point of the inference method is a representation of the system's behaviour as a string such as, for example, a time sequence of measurements of its location.[5] The symbols in this measurement sequence generally form a discrete and finite set (for background, see Section A in the Appendix). Once a string of measurement data has been obtained, the regularities are extracted using an algorithm which is briefly explained below, and a computational representation is inferred which reproduces the statistical regularities of the string. Computational representations can, in principle, be anything from the Chomsky hierarchy of computational devices (Hopcroft et al. 2001), but in concrete examples they usually are finite-state automata. The size of this automaton is the basis for the statistical complexity measure.

The algorithm for inferring the computational representation of a string assumes that a stationary stochastic process $\{X_t\}_{t \in \mathbb{Z}}$ has generated the string in question (for a definition of stationary stochastic process, see Section A of the Appendix). As a next step, statistically equivalent strings are grouped together. Two strings, $\underline{x}$ and $\underline{x}'$, are statistically equivalent if they have the same conditional probability distribution over the subsequent symbol $a \in \mathcal{X}$:

$$P(X_0 = a | X_t^{-1} = \underline{x}) = P(X_0 = a | X_{t'}^{-1} = \underline{x}'), \text{ for all } a \in \mathcal{X} . \quad (4.17)$$

The two strings do not have to be of the same length. The *equivalence class* of a substring $\underline{x}$ is denoted by $\varepsilon(\underline{x})$, and it contains all strings statistically equivalent to string $\underline{x}$, including $\underline{x}$ itself. These classes are called 'causal states', a somewhat unfortunate name since no causality is implied in any strict sense. Due to the stationarity of the process, the transition probabilities between the causal states are stationary and form a stochastic matrix. Hence, the computational representation obtained by this algorithm is a stochastic finite state automaton or, equivalently, a hidden Markov model (Hopcroft et al. 2001; Paz 1971) and is called ε-machine.[6]

The stationary probability distribution P of the ε-machine's causal states $s \in \mathcal{S}$, which is the left eigenvector of its stochastic transition matrix with

---

[5]Of course, measurement is crucial to computational mechanics, and it raises many practical questions left aside here.

[6]The inference algorithm is available in various languages, see, for example, Computational Mechanics Group (2015); Kelly et al. (2012); Shalizi and Klinkner (2003).

eigenvalue 1, is used to define the statistical complexity, C_μ, of a process:

$$C_\mu := - \sum_{s \in \mathscr{S}} P(s) \log_2 P(s) , \qquad (4.18)$$

where $\mathscr{S}$ is the set of causal states. C_μ is the Shannon entropy of the stationary probability distribution. This reflects the computational viewpoint of the authors since, technically, the Shannon entropy is the minimum number of bits required to encode the set $\mathscr{S}$ with probability distribution P. Thus, the statistical complexity is a measure of the minimum amount of memory required to optimally encode the set of behaviours of the complex system. It is worth noting that, for a given string, the statistical complexity is lower bounded by the excess entropy/predictive information (see eq. 4.11 above), $C_\mu \geq I_{pred}$ (eq. 4.11) (Crutchfield and Feldman 2003; Crutchfield et al. 2009; Wiesner et al. 2012). This mathematical fact agrees with the intuition that a system must store at least as much information as the structure it produces. The statistical complexity has been computed for the logistic map (see Chapter 1) (Crutchfield and Young 1989), for protein configurations (Kelly et al. 2012; Li et al. 2008), atmospheric turbulence (Palmer et al. 2000), and for self-organisation in cellular automata (Shalizi et al. 2004).

Crutchfield (1994, p. 24) writes that "an ideal random process has zero statistical complexity. At the other end of the spectrum, simple periodic processes have low statistical complexity. Complex processes arise between these extremes and are an amalgam of predictable and stochastic mechanisms." This statement, though intuitive, is obscuring the fact that the statistical complexity increases monotonically with the order of the string. For a proof consider the following. For a given number of causal states the statistical complexity has a unique maximum at uniform probability distribution over the states. This is achieved by a perfectly periodic sequence with period equal to the number of states. When deviations occur, the probability distribution will, in general, not be uniform anymore, and the Shannon entropy and with it the statistical complexity will decrease. On the other hand, increasing the period of the sequence requires an increased number of causal states and, thus, implies a higher statistical complexity. Hence, the statistical complexity scores higher for highly ordered strings than for strings with less order or with random bits inserted. The statistical complexity is a measure of order produced by the system, as well as a measure of memory of the system itself. The strength of the framework of computational mechanics lies in detecting order in the presence of disorder.

4.10.3 Effective Complexity

Effective complexity was introduced by physicists Murray Gell-Mann and Seth Lloyd (1996). Gell-Mann and Lloyd's starting point is common to many measures of complexity of that time: the measure should capture the property of a complex system of being neither completely ordered nor completely random. They assume the complex system can be represented as a string of bits; call it s. This string is some form of unique description of the system or of its behaviour or the order it produced. The algorithmic complexity (for a definition, see Section C in the Appendix) of this string of bits is a measure of its randomness or lack of compressibility. The more regularities a string has, the lower is its algorithmic complexity. Hence, Gell-Mann and Lloyd consider the algorithmic complexity not of the string itself, but of the ensemble (a term taken from statistical mechanics) of strings with the same regularities as the string in question. Let E be this ensemble of strings with the same regularities. The effective complexity of the string (and thus the system which it represents) is defined as the algorithmic complexity of the ensemble E in which it is embedded as a typical member. ('Typical' is a technical term here, but it captures exactly what we intuitively think 'typical' should mean.) Ensemble members $s \in E$ are called typical if $-\log \Pr(s) \approx K_{\mathcal{U}}(E)$, where $K_{\mathcal{U}}(E)$ is the algorithmic complexity of E (see Section C of the Appendix).[7] Assigning a probability to each string in a set is less arbitrary than it sounds. It has been shown that the probability $\Pr(s)$ of a string s is related to its algorithmic complexity as $-\log \Pr(s) \approx K_{\mathcal{U}}(s|E)$ where $K_{\mathcal{U}}(s|E)$ is the algorithmic complexity of the string s given a description of the set E. The effective complexity $\varepsilon(s)$ of a string s is defined as the algorithmic complexity of the ensemble E of which it is a typical member,

$$\varepsilon(s) = K_{\mathcal{U}}(E) . \tag{4.19}$$

For example, the ensemble of a string which is perfectly random is the set of all strings of the same length. This set allows for a very short description, by giving the length of the strings only. This trick of embedding the string in a set of similar strings exactly achieves what Gell-Mann and Lloyd set out to do. A string with many regularities over many different length scales, which is how we think of a complex system, will be assigned a high effective complexity. Random systems, in their structure or behaviour, will be assigned very low effective complexity. According to Gell-Mann and Lloyd (1996), the effective complexity can be high only in a region intermediate

---

[7]The idea of replacing entropy (the average of $-\log \Pr(s)$ is an entropy) by algorithmic complexity goes back to Wojciech Zurek (1989).

between total order and complete disorder. However, replacing some of the regular bits in a string by random bits decreases its regularities and hence its effective complexity. Just like the true measure complexity and the statistical complexity, the effective complexity increases monotonically with the amount of order present. This places the effective complexity among the measures of order. Gell-Mann and Lloyd note that this measure is subjective, since what to count as a regular feature is the observer's decision. The instruction "find the ensemble (of a rain forest, for example) and determine its typical members" leaves too many things unspecified for this measure to be practicable (McAllister 2003).

4.10.4 Logical Depth

Computer scientist Charles Bennett introduced *logical depth* to measure a system's history (Bennett 1991). Bennett argues that complex objects are those whose most plausible explanations involve long causal processes. This idea goes back to Herbert Simon's influential paper, 'The Architecture of Complexity' (1962). To develop a mathematical definition of causal histories of complex systems, Bennett replaces the system to be measured by a description of the system, given as a string of bits. This procedure should be very familiar by now. He equates the causal history of the system with the algorithmic complexity of the string (the length of the shortest program which outputs the string; see Section C of the Appendix). The shorter the program which outputs a system's description, the more plausible it is as its causal history. A program consisting of the string itself and the 'print' command has high algorithmic complexity, and it offers no explanation whatsoever. It is equivalent to saying 'It just happened' and so is effectively the null-hypothesis. A program with instructions for computing the string from some initial conditions, on the other hand, must contain some description of its history and thus is a more explanatory hypothesis.

In addition to considering a program's length as a measure of causal history, Bennett also takes the program's running time into account. A program which runs for a long time before outputting a result signifies that the string has a complicated order that needs unravelling. The definition of logical depth is then as follows. Let x be a finite string, and $K_{\mathscr{U}}(x)$ its algorithmic complexity. The logical depth of x at significance level s is defined as the least time $T(p)$ required for program p to compute x and then halt where the length of program p, $l(p)$, cannot differ from $K_U(x)$ by more than s bits,

$$\text{Depth}_s(x) := \min_p \{T(p) : l(p) - K_{\mathscr{U}}(x) \leq s , \ (U(p) = x)\} . \qquad (4.20)$$

Logical depth combines the features of order and history into a single measure. Consider the structure of a protein, for example. One possible program prints the electron and nuclear densities verbatim, with discretised positional information, which is a very long program running very fast. Another program computes the structure ab initio by running quantum chemical calculations. This would be a much shorter program but running for a long time. The latter captures the protein's order and history. The real causal history of a protein is, of course, very long, starting with the beginning of life on Earth, or even with the beginning of the universe. The logical depth captures our intuition that complex systems have a long history. A practical problem is that the time point when the history of a system starts is not well defined. Another aspect which makes it impractical to use is that the algorithmic complexity is uncomputable in principle, although approximations exist such as the Lempel-Ziv algorithm (Ziv and Lempel 1977). The next chapter returns to the question of 'What is a complex system?'

Chapter 5

What Is a Complex System?

Is there a single natural phenomenon of complexity found in a wide variety of living and nonliving systems and which can be the subject of a single scientific theory? Is there such a thing as 'complexity science' rather than merely branches of different sciences, each of which have to deal with their own examples of complex systems? This chapter synthesises an account of how to think about complexity and complex systems from the examples and analysis of the preceding chapters. Roughly speaking, our answers to these questions are no to the first and yes to the second. There is no single phenomenon of complexity, but there are a variety of features of complex systems that manifest themselves in different ways in different contexts. Hence, complexity science is not a single scientific theory but a collection of models and theories that can be used to study the different features in common ways across different kinds of systems. The following sections consider different views about complex systems, and the penultimate section argues for our view. The final section of this chapter reflects on the broader implications of what has been learned.

5.1 Nihilism about Complex Systems

The most negative view of complexity is simply that there is no such thing. Similarly, it could be argued that 'complex system' is a vague and ambiguous term that covers a variety of things and that complexity science is just a collection of techniques and methods that does not have a domain of its own. In this way of thinking, which we can call 'nihilism', there are very different ways that different sciences are combined to study complicated systems, and complex systems are just complicated things we study with interdisciplinary science using computers. Anything over and above that is, at best, a convenient label for an emerging synthesis (in the sense discussed in Chapter 1) and, at worst, hype for the purposes of grant applications, journal publications and book sales. On this kind of view, it seems advisable to stop using the terms 'complexity' and 'complex system' as if they were well-defined scientific concepts.

Consider nihilism in the light of the discussion in Chapter 2. For example, take the human brain. It is ultimately subject to the laws of physics, chemistry and biochemistry. All the basic processes that occur in it can be described in the language of these existing scientific disciplines. It might be argued that the necessity to describe the brain as a complex system, and to use the new techniques of complexity science, is simply due to the fact that the physical and chemical mechanisms relevant for its behaviour are so numerous and complicated that it is impossible to describe them all in practice. Once one starts to care about details, underlying mechanisms and scientific explanations of them, the brain necessarily enters the realm of physics, chemistry and biochemistry. So described, the brain is the subject of each of the traditional sciences, and it might be supposed it could be completely described in the language of existing physics and chemistry. Unless there is a fundamental new set of forces and particles associated with the brain, for which there is no evidence, the brain is a physical and biochemical system. Theories of the brain ultimately rely on basic science and fit into the hierarchical organisation of nature above physics, chemistry and biology. Similarly, for the rest of the examples so it might be argued, there is no single concept of complex system; there are just systems that are physical, chemical, biological, economic, or whatever.

Of course, nihilists must admit that the complex systems perspective has allowed problems to be discovered and discussed in an unconventional, interdisciplinary way. For example, scientists now study the immune system as a network of nodes and links. However, they do not necessarily regard this as a new science, because this perspective is embedded into the existing frameworks of anatomy and physiology. In this view, there are two possible paths that a new problem and its solution can take in the scientific disciplinary landscape: either it can be completely swallowed by the existing field as if it had always been part of it, or it can become a field in its own right, such as neuroscience, which uses ideas from network theory. Either way, the problems and ansatz to solutions only remain part of the field of complex systems until the system has been fully understood. According to nihilism, the field of complexity science is just a feeder of new science to other fields but not a science in its own right.

This view does not directly contradict any of the truisms of Chapter 1, but it does not take account of the fact that there are kinds of invariance and universal behaviour studied by complexity science that are found in many very different kinds of systems. Furthermore, as shown in the last chapter, there are novel kinds of invariance and forms of universal behaviour that are found when complex systems are modelled as networks and information-processing systems. Hence, nihilism about complex systems should be rejected.

5.2 Pragmatism about Complex Systems

A different view agrees with nihilism that there is no essence of complexity or complex systems but is nonetheless much more positive about complexity science. 'Pragmatism' says that there is enough of a resemblance between the collection of methodologies, mathematical and computational, which are applicable to a whole variety of systems, from proteins to the brain to forests and cities, in order for the ideas of complexity and complex systems to be scientifically useful. In this approach, the term 'complex system' refers to a collection of systems in various disciplines that are all amenable to related mathematical and computational techniques, such as network theory or agent-based modelling, but which otherwise have nothing fundamental in common. According to this approach, the wide applicability of complexity science is not necessarily indicative of any unity to the idea of a complex system. The techniques are useful to scientists, so they are worth teaching to students, and this is sufficient to justify the existence of the many research centres for complexity science throughout the world. Pragmatism says complexity science is a range of models, theories and techniques that are applicable very widely. Strevens (2016) advocates pragmatism about complexity science in this sense.

Pragmatism is probably a common view among complexity scientists who have no interest in worrying about what complexity is or what complex systems are.[1] However, there are many who go further and argue for particular definitions of complexity and complex systems or at least who impose necessary conditions that rule out some cases. Some of these conceptions of complexity are considered below (they all originate with scientists working in the field). The pragmatist about complex systems need not deny any of these views but may be agnostic about them.

---

[1] Similarly, computer scientists do not need to worry about what a computer is. However, Alan Turing's investigations into the latter question led to the theory of Turing machines, which is a valuable part of computer science.

5.3   Realism about Complex Systems

Realism about complex systems is the idea that 'complex systems' form what philosophers call a 'natural kind'. Most scientists probably think that the elements of the periodic table, or the fundamental particles that make them up, are the ultimate natural kinds. In science, it is crucial to find the right way of dividing things up into kinds. For example, whales should not be put in the same category as fish just because they live in the sea and have fins, and jade is not in fact a natural kind of mineral (because there are two different minerals, jadeite and nephrite, that have both been given the same name). Every science has its own taxonomy. There are white dwarfs and red giants in astrophysics, noble gases and halogens in chemistry, and organisms and classes of them in biology. Often the way things are classified is refined over time, and arriving at an exact definition may be difficult or impossible. For example, the definition of an acid has undergone various revisions, and there is no consensus about what exactly life is and whether viruses are alive. This implies that if the concept 'complex system' is vague, that does not make it any less scientifically valid than concepts such as that of life. The realist can argue that just as living systems form a natural kind, so too do complex systems, but only if they are able to say something, vague or not, about what complex systems are.

It is important to realism about science in general that there are natural properties that are independent of our conventions and interests. For example, the mass of objects or their electric charge is what it is, however we decide to quantify them with a choice of units. One version of the realist view of complex systems, popular with many complexity scientists, is that there is such a natural property of complexity that can be quantified and measured. Then complex systems are derivatively just those that have a lot of complexity.

As discussed at the end of Chapter 1, scientists who hold the realist view are not at all a uniform group. Some understand complexity narrowly so that only biological systems count as complex. Others – for example, Murray Gell-Mann (1994) – understand it more broadly so that physical and chemical systems that are not associated with life can also be complex. As a result, among the realist conceptions of complexity that are proposed in the literature, some are couched in generic scientific language, some involve only physical quantities, some use computational/information-theoretic concepts, and some require biological ideas. Below, a representative sample of views is outlined.

5.3.1 Generic Conceptions of Complexity

Generic conceptions of complexity are those that can be stated using generic scientific or mathematical language (often that of dynamical systems theory). In particular, they do not make reference to physical quantities, computational or information-theoretic ideas, or biological concepts.

Complexity as Emergence

As defined in Chapters 1 and 2, the most basic kind of emergence is the existence of laws and properties at the level of the whole system that do not exist at the level of the constituent parts. As stated at the end of Chapter 3, the simplest realist view is that complex systems are all and only those that exhibit emergence. This view reflects the fact that everyone in the complexity science community takes emergence to be central and has done so ever since the first meetings in Santa Fe, as described in Chapter 1. This view is clear and can be made exact, because it is possible to describe mathematically how a system with very many degrees of freedom can have a much simpler lower-dimensional dynamics defined on effective states, as in statistical mechanics.

Emergence can be purely epistemological, meaning the emergent entities do not really exist but are just convenient ways to keep track of what is happening in the system. For example, the Game of Life shows a merely epistemological kind of emergence because 'eaters' and 'gliders' do not really eat or glide respectively; they are just a kind of useful fiction for keeping track of the evolution of the system. On the other hand, the emergence from the physical systems in the universe to the immensely intricate and structured system of life on Earth, including the human brain and the complexity of human culture and social life, is ontological emergence (unless there are not really animals and people and so on).[2]

Certainly emergence in all epistemological senses is necessary for a complex system. If a system does not exhibit higher-level order of some kind that arises spontaneously from the interactions of its parts, then it is not complex. However, as discussed in Section 2.1, even an isolated gas at thermal equilibrium has emergent properties of pressure and temperature that an individual molecule does not. Such a gas lacks all the interesting features of complex systems described in detail in Chapters 3 and 4 other than this most minimal kind of emergence, which is exhibited by everything in the sciences except perhaps the entities of the ultimate fundamental physics if it exists.

---

[2]Some philosophers deny that there is ontological emergence and claim that anything other than fundamental physical stuff is at best a kind of second-class thing (see Ney 2014, pp. 111–112). We reject this view.

Since there are different kinds of emergence, it may be argued that some specific kind of emergence is required of a complex system. This view may then coincide with one of the other views discussed below, depending on what kind of emergence is taken to be necessary.

Complexity between Order and Chaos

Many people say that a system is complex if it lies between order and chaos (Crutchfield 2012). As argued in Chapter 3, the notions of order and disorder are crucial to the understanding of complex systems, but they are not sufficient for a definition. The idea of complexity as some kind of middle ground between order and disorder is also misleading. A crystal with many impurities, which is a very ordered arrangement of atoms or molecules with some atoms slightly misplaced or replaced by different atoms, is not thereby more complex than a crystal with fewer impurities or more complex that a crystal with more impurities. Accounts of complexity based on 'measures of complexity' that measure order in one way or another confuse order with the way that it is produced, as argued in Chapter 4. Furthermore, as argued in Chapter 3, ideas of order and disorder are always relative to some way of representing the system or gathering data about it, and what is ordered at one scale may be disordered at another and vice versa. The analysis of the previous chapters shows that complexity is too diverse to be captured in this way.

5.3.2 Physical Conceptions of Complexity

Physical conceptions of complexity are those that use only the language of physics to characterise complexity. In particular, they do not make reference to computational or information-theoretic ideas or to biological concepts. Note that biological systems can count as complex systems (and could even be the only ones there are) according to physical conceptions of complexity. The important thing is that the following conceptions of complexity do not define it in terms of biological concepts, even if only living things satisfy such a definition.

Complexity as Non-Equilibrium Thermodynamics

Various authors characterise complexity in terms of thermodynamic properties such as entropy, energy and being out of equilibrium (see Chapter 3). For example, astrobiologist Charles Lineweaver (2013) explores the idea that the creation of complexity is driven by free energy and argues that, as the latter decreases with the increase of entropy, so too will complexity decrease as we head towards the heat death of the universe. The association of complexity with free energy is illustrated by the fact that the degree of order of a hurricane depends on pressure, temperature and humidity gradients that correspond to the gradient of free energy (recall also the discussion of Section 3.4. The formation and maintenance of biological structure also seems to require a free energy gradient. On the other hand, the astrophysicist Eric Chaisson (2013) proposes energy flow per unit time per unit mass as a measure of complexity. He applies this idea to both nonliving and living systems to argue that in both cases complexity has increased over time in the history of the universe. David Wolpert (2013) offers a way to use the Second Law of Thermodynamics to explain complexity by measuring it with a notion of self-similarity at different scales.

The relationships between thermodynamic concepts, such as entropy and free energy, and complexity are tantalising, but so far there is no agreement about what exactly they are. The advantage of such views is that there does indeed seem to be an important connection between free energy and complexity in many systems and that real complex systems are always open systems (as discussed in Chapters 2 and 3) (Smith and Morowitz 2016). One disadvantage is that the notion of free energy is defined in terms of the thermodynamic concepts of entropy and heat, which are well-defined in equilibrium thermodynamics only. The analysis of the preceding chapters shows that being open or driven is not sufficient for all the features of complex systems.

Complexity as Self-Organised Criticality

According to Per Bak (1996) the concept of self-organised criticality could be taken as the basis for a theory of complexity. The sandpile model shows that complex behaviour in the form of avalanches and the formation of structure can have universal aspects that arise spontaneously from simple interactions of many parts. Such self-organisation occurs only in open systems. It is a kind of emergence not exhibited by gases and condensed matter away from critical points but which falls well short of the more sophisticated features of complex systems involving various forms of adaptive behaviour. Hence, while it is an important feature of complex systems (see Chapter 4), it is not the only one.

5.3.3 Computational Conceptions of Complexity

Recall that the influential idea of 'logical depth' due to computer scientist Charles Bennett (1991) is based on the mathematical theory of computational complexity, which measures the degree to which strings of symbols depart from randomness (as is explained in Chapter 4). Physicist Seth Lloyd (2006) argues that complexity results from quantum computation, assuming both the existence of quantum multiverses and that complexity is measured by the notion of thermodynamic depth. Many other computational and information-theoretic measures of complexity have been proposed. Chapter 4 shows they measure not complexity but order. There is a rivalry between physical and computational and information-theoretic measures of complexity throughout debates about complexity science. The next section is about conceptions of complexity that one way or another involve the idea of function.

5.3.4 Functional Conceptions of Complexity

Chief among the examples of complex systems discussed in this book and more generally are living organisms, their parts, and collections of them – for example, cells, eusocial insect colonies, ecosystems, the immune system and the human brain. In the context of biology, complexity involves not only structure and its formation and maintenance, but also adaptive behaviour. As argued in Chapter 3, adaptive behaviour generates new forms of features of complex systems such as robustness and nested structure. For this reason, as mentioned in Chapter 1 and at the end of Chapter 3, many of those who have addressed the question of what is a complex system say that complex systems are those that display adaptive behaviour. Prominent examples include Mitchell (2011) and Holland (2014) (and this view is represented in the quote from Pines in Chapter 1). Biological complexity is central to Herbert Simon's famous idea of the hierarchy of complexity discussed in Chapter 3.

However, the line between the living and nonliving is not the same as the line between things that display adaptive behaviour and those that do not. Adaptation is always relative to some notion of function, whether this be simply the goal of reproduction or a derivative goal like navigation. All living systems can be understood in functional terms; however, there are also nonliving complex systems that have inherited goals and purposes from us. Adaptation exists in artificial systems that are not alive, such as in software. The complex systems of human construction such as markets and economies, IT networks, transportation networks and cities can all be thought of in functional terms – for example, Internet routers that optimise the flow of data. Much of complexity science is the study of adaptive behaviour in living systems or in those we have engineered, and it could be argued that all genuinely complex systems are either living systems or systems created to serve the purposes of living systems. Nonetheless, the relevant concept is not that of life but that of function (or relatedly, goal or purpose).

It may seem obvious that, given the difference between earthworms and human beings, biological complexity has increased more or less monotonically during the course of the history of life on Earth; however Stephen Jay Gould argues that this is based on a very simplistic understanding of evolution (see, for example, Gould 2011). Although biological complexity is much discussed by influential figures such as Simon Conway-Morris and Stuart Kauffman, there is no general agreement about what it is. Proposed measures of it include genome length and the subtler notion of generative entrenchment. Like the idea of logical depth mentioned above, biological complexity is often related to the history of a system or its relation to its environment. Physicist Eric Smith (2013) analyses biological complexity in terms of the notions of memory and robustness that often figure in the science of complex systems. A flock of birds dispersed by a predator reforms and returns to its trajectory, despite lacking any overall controller. Smith connects this with the formalism of order parameters in statistical physics mentioned in Chapter 2. On the other hand, biologist David Krakauer (2013) thinks of biological complexity in terms of the cognitive capacity of an organism to represent and predict its environment.

None of the above accounts of complexity and complex systems is complete. Biological conceptions do not apply to nonliving systems, and functional conceptions more generally do not apply to systems that lack function, goals or purposes like the BZ reaction and the solar system. However, as shown in the previous chapters, such nonliving systems exhibit many of the features of complex systems. None of the other conceptions of complexity discussed above are satisfactory on their own, because they either depend on the speculative extension of thermodynamic or computational/information-theoretic concepts beyond their domains of application or are based on measures of one aspect of complex systems, such as the order that they produce or their robustness, while neglecting others. Chapter 4 discussed in detail some of the many measures of complexity that have been proposed. If we are right that none of them measures complexity as such, but rather features of complex systems, then realism about complex systems based on any of them is not viable. In the next section we explain our answer to the question 'What is a complex system?' The view advocated may be thought of as a version of realism that takes the conceptions of complexity above to be expressing different features of complex systems so that there are different kinds of complex systems, both because not all have all the features and because the different features take different forms.

5.4 The Varieties of Complex Systems

Complexity science studies how real systems behave. The models of the traditional sciences often treat systems as closed. Real complex systems interact with an environment and have histories. Complexity is not a single phenomenon but the features of complex systems identified in Chapter 3 are common to many systems. If it is right that the hallmark of complex systems is emergence and that there are different kinds of emergent features of complex systems, then instead of defining a complex system in terms of one particular kind of emergence, it is possible to identify different varieties of complex systems according to what emergent features they exemplify. This view is explained below, and the most important lessons of this book are revisited and brought together.

The truisms of Chapter 1 fit with our analysis. As explained at the end of Chapter 1, complexity science began with the search for new syntheses between models and theories from different domains. For this reason it involves multiple disciplines. Complexity science is possible because despite the differences between different complex systems, there are kinds of invariance and forms of universal behaviour in complex systems (truism 5), and many of these can be described mathematically, as shown in Chapter 4. The new kinds of universality and invariance that emerge when complex systems are modelled as networks and information-processing systems include scaling laws and some of the forms of robustness and modularity explained in Chapter 4.

Simple and regular behaviour of the whole can emerge from very complicated and messy underlying behaviour of the parts, as with the dynamics of some physical systems like the periodicity of a chemical oscillator. On the other hand, the adaptive behaviour of an ant colony can be generated by relatively simple rules governing individuals. Clearly, 'simplicity' means many different things in the context of complex systems, but the features of complex systems can arise from relative simplicity because of the numerosity of interactions and external drivers.

Complexity science is probabilistic in part because it uses ideas and techniques from statistical mechanics, and in part because it models systems in terms of information and order which is represented with probability theory. Complexity science is computational in two ways. First, it uses computational models of systems such as condensed matter, the solar system, the climate and so on. Second, it models systems as themselves performing computations. This requires a functional understanding of the system.

Complexity science does study something distinctive – namely the emergent features of systems that are composed of a lot of components that interact repeatedly in a disordered way. The reason why it has been hard to identify what is distinctive about complex systems is that there are many different kinds of emergent properties and products of complex systems, and they are not all found in all complex systems. The common features of complex systems manifest themselves differently in different kinds of systems. For example, the robustness of an insect colony is different from the robustness of a weather pattern. It is essential also to distinguish the order that complex systems produce and the order of complex systems themselves.

Emergence often takes the form of relatively sharp boundaries in time, space or both. The examples from the physical world show that these sharp boundaries (like thermoclines and the heliopause in the solar system) and emergent entities and properties (like gases with their pressure) are relative to scales of energy, space and time. Section 2.1 shows that much of physics is about emergent phenomena and that the chemistry and physics of nonliving systems exemplifies all of the other truisms of complexity science. The fact that nonliving systems can generate order is exemplified by matter, the Earth, and the solar system.

Recall that order, structure or form is inhomogeneity of some kind. Hence, in general, order is associated with symmetry breaking of some kind. There is a general idea of equilibrium states as states that do not change in some relevant respect over some relevant time scales. In some living systems there is a very narrow range of temperature that allows for the stability of the physical processes that sustain life. Phase transitions are emergent order in the dynamics of systems of many parts, and they may also produce structure in the systems with which they interact. Both symmetry breaking and phase transitions are found throughout physics from cosmology to quantum field theory. Universal aspects to critical phenomena are found in many different contexts and in many very different kinds of processes.

An isolated gas at thermal equilibrium does not exhibit any feedback or self-organisation, and it does not have any of the products of complex systems other than the emergence of the most basic kind of order (in the form of the ideal gas laws). All complex systems that have any kind of self-organisation beyond this are open to the environment and driven from outside in some way. Driven physical systems like condensed matter or gases undergoing heating across a critical point exhibit all the rich structure of phase transitions, and these involve highly nonlinear dependencies between their degrees of freedom.

The related phenomenon of a simple physical system self-organising close to but without crossing a critical point has led to the term 'self-organised criticality', which was discussed above and in Chapter 4. The canonical example of this phenomenon is the pile of sand, which originally was only a computer algorithm. While systems at phase transitions exhibit order and nonlinearity (in the form of power-law behaviour), self-organised critical systems represent a higher level of complexity since they also exhibit robustness as they stay close to a critical point under a range of external conditions exhibiting correlations on many length scales. Another example of a driven system is the BZ reaction, which spontaneously produces order and obeys nonlinear emergent dynamics. These are complex systems of the most simple kind. A much bigger and more complex purely physical system like the solar system, on the other hand, has further features, including history and nested structure, as well as robustness of various kinds.

Clearly, more is different in many different ways, and the 'more' in question can be parts, interactions, connections, and so on. There are many different kinds of emergence, and not all complex systems exemplify all of them. The full extent of the emergence of the world around us did not happen overnight. Even the most simple nonliving things with which we interact, such as stones and the wind, exist only because of a very long and complex history. This is even more true for living beings, which carry a lot of the history of life within them, as well as the history of the matter of which they are made. In multi-cellular life in general, and especially in humans and our brains, function is built on function so that history is encoded in both mechanisms and structure.

The examples from the physical world are sufficient to show that there are many forms of structure, at many different length and time scales and that there are many examples of relatively separate slow and fast dynamics within a single system in physics. Processes on different time scales can become effectively decoupled. Indeed, physics is full of precise laws about relatively isolated systems and their different kinds of properties at many different scales because of the effective decoupling of scales. Biology is not like this because processes on very different scales are tightly coupled in organisms.

Decoupled fast and slow dynamics can produce and maintain the stability of structure, as within atoms and the solar system. The climate is a case where the structure of the system and the structure produced by the system are intricately related. In the atmosphere there are interconnected processes on many different length and time scales and with both positive and negative feedback loops. The climate and the economy illustrate how complex systems can go wrong when processes at different scales become tightly coupled. Complex systems often have very different equilibria in which the very same collection of parts displays very different emergent properties. For example, a stable weather pattern in the form of seasonal rains can disappear, and stable prices can become hyperinflation.

In general, positive feedback is often destabilising and negative feedback stabilising. The discussion of markets and economies shows that the effects of feedback can be highly unpredictable. In general, the role of feedback in the interaction between people and technology is extremely important. For example, positive feedback loops have led to the dominance of very large service providers and to people mediating more and more of their interactions with each other by smartphones. The next most important feedback effect will be between human beings and intelligent software agents and big data (Burr et al. 2018).

It is stochasticity and feedback that generate higher-level approximate order. Only when there are many instances of probabilistic processes does average behaviour dominate what happens. Hence, the kind of robustness to perturbation found in complex systems requires large numbers. This is another way 'more is different'. Larger numbers are needed for interactions between parts to become frequent enough for self-organisation to occur. This is yet another variation on 'more is different'. It is important that the idea of interaction here is the general one of *dependence* of the states of the elements on each other, which may be mediated by different kinds of physical interactions.

As noted in Chapter 1 and above, the disagreement about how to define complexity goes along with disagreement about which systems are complex. Nobody denies the brain is a complex system, but the solar system is more contentious. Certainly, the brain is the complex system that exhibits all features of complexity to the highest degree. However, to say that the BZ reaction is not a complex system but the brain is misses the point that they share the basic conditions of complex systems and some of the important products, even though they do not both have all of the latter.

Complexity is a multi-faceted phenomenon that has a variety of features not all of which are found in all complex systems but which are related, as when feedback produces nonlinearity. The 'complexity' measures applicable to real-world complex systems do not measure complexity as such but rather a variety of features, each of which can take different forms. Hence, rather than restricting the term 'complex system' to systems displaying adaptive behaviour or using it completely generally as if all complex systems were simply those displaying some kind of self-organisation, our view is that there are different kinds of complex systems and that different features of complexity are displayed by them. The 'conditions' for complexity – numerosity, disorder and diversity, feedback, and non-equilibrium – give rise to the 'products' of order and organisation, robustness, and nonlinearity. Figure 5.1 shows a sketch of these features of complexity. Systems with all the features of order, robustness, nonlinearity, nested structure and history are at the highest level of complexity found in systems that are not ascribed functions. Once function is involved then adaptive behaviour leads to the functional features of forms of self-organisation and order, modularity, and memory – which are interlinked and which require all of the other conditions and products.

However, there are also more or less sophisticated forms of adaptive behaviour, and one might assign a degree of adaptive behaviour to the Earth's climate. This explains why what people mean by 'complexity' and 'complex system' is so fluid. They are sometimes talking about some or all of the products, sometimes about the conditions, and sometimes about both. The climate is the biggest system we know of at that level of complexity. But the climate also illustrates how permeable the boundaries are between the different levels of complexity. The Earth's atmosphere developed because of the simultaneous evolution of microbes, exemplifying the intricate link between living and nonliving matter. One might even argue that the climate is performing a function, that of making life on Earth possible in its synergy between living organisms and lifeless matter. (The Gaia hypothesis makes this view explicit; Lovelock and Margulis 1974.)

It is fundamental to all complex systems that have all the features above that they have a very long history and that their interesting features are apparent only at the right spatial and temporal scales. The stability of complex systems arises spontaneously only under the right conditions. If they are perturbed too much, the products of them that we value can dissolve into the disorder that underlies them. Interference with them can shift them into new states or into states in which complexity no longer emerges.

In answer to the questions at the beginning of the preface, complexity is not just a label, but it is not a single phenomenon. The different conceptions of complexity of physicists, biologists, social scientists and others cannot be brought into a single framework, because the features of complex systems are many and take different forms. However, there are multiple frameworks, such as information theory, network theory, and the theory of critical phenomena, that can be applied across the sciences. Measures of complexity are meaningful but measure different features of complex systems that manifest themselves in different ways.

Figure 5.1: The products of complexity, separated into those that appear in matter, functional systems and living systems and those that appear in functional and living systems only.

5.5 Implications

There are two broad kinds of implications of this book – namely the philosophical and the practical – and they shade into one another when the ethics and politics of complexity science are considered. There are many important theoretical questions on which complexity science bears, the most obvious ones concerned with the relationships between life and nonliving matter, and between conscious and non-conscious matter. The general implication of our analysis for these matters is that the dichotomy between atoms and molecules and advanced life forms is a very crude way of seeing the many layers of structure that are found at different scales. The only way to understand the emergence of life is by studying the processes that occur in self-organising physical systems not just physical structures. Once the complexity of nonliving systems, such as the solar system and the Earth and its climate, is grasped in detail, the difference between life and non-life seems to be less of a mysterious leap and more of a continuum. Similarly, the intelligent collective behaviour of eusocial insects and the many layers of complexity between the simplest nervous systems and the human brain make the consciousness of people (who of course have been subject to the very many interactions of development) seem like a very complicated bundle of capacities, not a single property. More generally, one of the main metaphysical lessons of this book is that emergence comes in many different forms at many different scales and should be understood in terms of interactions and processes.

One of the most important practical goals of complexity science is to improve the functioning of societal processes that make important decisions. An important lesson of complexity science is that complex systems can spontaneously produce emergent phenomena in some regimes but not in others. There are many social processes at work in the scientific community, which itself may be thought of as a complex system. As such, at its best, it has the emergent properties of error correction and truth tracking and produces incredibly accurate and precise information about the world. However, complex systems can be pushed into very different equilibria in which they completely lose some or all of their emergent properties. Care must be taken to ensure that the incentives and forms of interactions that scientists are subject to continue to produce science as a collective result. Clearly, the complex systems of the climate and ecosystems may be destabilised to the point of radical change due to the coupling of processes at different time scales and to various kinds of perturbations that may not be obviously significant in advance. The dramatic changes currently being witnessed may be only the tip of the iceberg.

Recall the example of the army ants and the need for sufficient numbers for their interactions to produce collective behaviour. Hence, collective human behaviour may be similarly sensitive to the nature and number of interactions between individuals. As technology is increasingly eliminating or mediating these interactions, it is possible that human beings and human society will change radically in ways that cannot be anticipated. Positive feedback loops can act very forcefully and fast, as anyone who has heard the sound from a microphone fed back into it from loudspeakers knows. The sound volume goes up increasingly fast until someone or something cuts the connection. There are many positive feedback loops operative at the moment as the use of different technologies goes from being exotic to optional to effectively compulsory for individuals.

The fields of AI and Big Data are about complex systems. The old way to think about artificial intelligence involved a computer program that was designed in advance and didn't change when in use. The systems that are now used to solve the classic problems such as face recognition and translation are open systems that have interacted extensively before they are used and continue to do so once they are in use. It is arguable that new complex systems are being created whose emergent behaviour is completely unpredictable and some of which may not be desirable. For example, the complex system that consists of social media users and the algorithms that select the content that is most circulated have created emergent phenomena such as 'trending' and can allow antisocial attitudes and misinformation to propagate more effectively that prosocial attitudes or facts respectively.

In general, since the equilibria that dominate the natural and social worlds have evolved over millennia to be compatible with human flourishing, it is much more likely that any new equilibria into which they shift will be to our detriment, and it is possible that they will lead to the destruction of much or all of what we value. The universe seems to have become more and more complicated since its infancy, and the most intricate structures in the universe of which we know (which are living systems) seem to be more complex than they have ever been. However, there is no guarantee that this will continue, and the fate of complex systems such as the climate, the economy and life on Earth depends on us.

Appendix – Some Mathematical Background

In this Appendix, all mathematical terminology used in the main text is defined, and some more background is given to the mathematical formalisms. The areas which are included are probability theory, information theory, algorithmic complexity, and network theory. All terminology which is defined is typeset in bold.

A   Probability Theory

An **alphabet** $\mathscr{X}$ is a set of symbols, numeric or symbolic, continuous or discrete, finite or infinite. Symbolic alphabets are discrete; continuous alphabets are numeric and infinite. $|\mathscr{X}|$ denotes the size of set $\mathscr{X}$. An example of a numeric finite, discrete alphabet is the binary set $\{0,1\}$; an example of a symbolic, finite alphabet is a set of Roman letters $\{a,b,c,\ldots,z\}$. An example of a numeric infinite, discrete alphabet is that of the natural numbers $\mathbb{N}$, an example of a continuous alphabet is the set of real numbers $\mathbb{R}$. This book discusses only discrete alphabets.

A **discrete random variable** X is a discrete alphabet $\mathscr{X}$ equipped with a probability distribution $P(X) \equiv \{\mathrm{Pr}(X=x),\ x \in \mathscr{X}\}$. We denote the probabilities $\mathrm{Pr}(X=x)$ by $P(x)$ or sometimes, to avoid confusion, by $P_X(x)$. The **uniform distribution** of a set $\mathscr{X}$ is the distribution $P(x) = 1/|\mathscr{X}|$ for all $x \in \mathscr{X}$. For two discrete random variables, X and Y, the joint probabilities $\mathrm{Pr}(X=x, Y=y)$ on alphabet $\mathscr{X} \times \mathscr{Y}$ are denoted by $P(xy)$ or sometimes, to avoid confusion, by $P_{XY}(xy)$. The joint probability distribution induces a conditional probability distribution $P(x|y) \equiv \mathrm{Pr}(X=x|Y=y)$, which is a probability distribution on $\mathscr{X}$ conditioned on Y taking particular value $Y=y$. Any joint probability $P(xy)$ can be written as

$$P(xy) = P(x|y)P(y) . \tag{1}$$

The **expectation value** of a discrete numeric random variable X, denoted by $\langle X \rangle$, is defined as

$$\langle X \rangle := \sum_{x \in \mathscr{X}} P(x)x . \tag{2}$$

Another common notation for the expectation value of X is $\mathbb{E}X$.

The **variance** of a numeric random variable X is the average deviation of X from its expectation value. Denoted by $\operatorname{Var}X$, it is defined as

$$\operatorname{Var}X := \langle (X - \langle X \rangle)^2 \rangle . \tag{3}$$

The square root of the variance of a random variable X is called the **standard deviation** σ:

$$\sigma = \sqrt{\operatorname{Var}X} . \tag{4}$$

The ratio of variance to expectation value is called the **coefficient of variation**,

$$\mathrm{cv} := \frac{\sqrt{\operatorname{Var}X}}{\langle X \rangle} . \tag{5}$$

The **covariance** of two numeric random variables X and Y is defined as

$$\begin{aligned} \operatorname{Cov}XY &:= \langle (X - \langle X \rangle)(Y - \langle Y \rangle) \rangle \\ &= \langle XY \rangle - \langle X \rangle \langle Y \rangle . \end{aligned} \tag{6}$$

A **stochastic process** $\{X_t\}_{t \in T}$ is a sequence of random variables X_t, defined on a joint probability space, taking values in a common set $\mathscr{X}$, indexed by a set T which is often $\mathbb{N}$ or $\mathbb{Z}$ and thought of as time. This book only discusses discrete time processes. A stochastic process is called a **Markov chain** if X_t (sometimes called 'the future') is probabilistically independent of $X_0 \ldots X_{t-2}$ ('the past'), given X_{n-1} ('the present'); in other words,

$$P(X_t|X_0 \ldots X_{t-1}) = P(X_t|X_{t-1}) , \text{ for all } t \in T . \tag{7}$$

A stochastic process is **stationary** if

$$P(X_t X_{t+1} \ldots X_{t+m}) = P(X_{t'} X_{t'+1} \ldots X_{t'+m}), \text{ for all } t, t' \in T, m \in \mathbb{N}. \tag{8}$$

A **hidden Markov model** $\{X_t, Y_t\}_{t \in T}$ is a stationary stochastic process of two random variables X_t and Y_t which forms a Markov chain in the sense that Y_t depends only on X_t, and X_t depends only on Y_{t-1} and X_{t-1}:

$$P(Y_t|X_0 \ldots X_t Y_0 \ldots Y_{t-1}) = P(Y_t|X_t) \qquad\qquad \text{and} \tag{9}$$
$$P(X_t|X_0 \ldots X_{t-1} Y_0 \ldots Y_{t-1}) = P(X_t|X_{t-1} Y_{t-1}) , \qquad \text{for all } t \in T . \tag{10}$$

136

The graphical representation of a hidden Markov model is a directed graph where the states are the outcomes $x \in \mathcal{X}$ of the random variable X_t and the state transitions are labelled by the outcomes $y \in \mathcal{Y}$ of the random variable Y_t and the corresponding conditional probability $P(Y_{t+1} = y, X_{t+1} = x | X_t = x)$.

B Shannon Information Theory

In the 1940s, the American engineer Claude Shannon, working for Bell Labs, introduced a mathematical theory of communication that is now at the heart of every digital communication protocol and technology, from mobile phones to e-mail encryption services and wireless networks (Shannon 1948). Shannon was concerned with defining and measuring the amount of information communicated by a message transmitted over a noisy telegraph line. He saw a message as communicating information if the receiver of the message could not predict with certainty which message out of a set of possible ones she would receive. By setting the amount of information communicated by a message x as proportional to its inverse log probability $1/\log P(x)$, Shannon axiomatically derived a measure of information, now called Shannon entropy. The **Shannon entropy**, a function of a probability distribution P but often written as a function of a random variable X, is defined as follows (Cover and Thomas 2006):

$$H(X) := - \sum_{x \in \mathcal{X}} P(x) \log P(x) , \tag{11}$$

where the log is usually base 2 and $0 \log 0 := 0$. The equivalent definition

$$H(P) := - \sum_{i=1}^{n} p_i \log p_i , \tag{12}$$

where $P = \{p_1, p_2, \ldots, p_n\}$, makes it explicit that H is a function of the probabilities alone, independent of the alphabet $\mathcal{X}$. This book discusses only the entropy of finite probability distributions, but the definition of the Shannon entropy extends to infinite but discrete, as well as to continuous probability distributions. Taking the logarithm to base 2 is a convention dating back to Shannon, due to a *bit* being the essential unit of computation. For a given set of messages $\mathcal{X}$, the Shannon entropy is maximum for the uniform distribution and proportional to the logarithm of the total number of messages. This illustrates that the Shannon entropy is a measure of randomness. If one of the messages has probability 1 and the others have probability 0, then the message is perfectly predictable, and the Shannon entropy is zero. The Shannon entropy is also precisely the expectation value of the function $1/\log P(x)$.

The **joint entropy** of n random variables $X_1, \ldots, X_n$ with joint probability distribution $P(X_1 X_2 \ldots X_n)$ is defined as

$$H(X_1 X_2 \ldots X_n) := - \sum_{\substack{x_1 \ldots x_n \in \\ \mathscr{X}_1 \times \cdots \times \mathscr{X}_n}} P(x_1 x_2 \ldots x_n) \log P(x_1 x_2 \ldots x_n) . \qquad (13)$$

Consider two random variables X and Y and joint probability distribution P_{XY}. The **conditional entropy** of X given Y, $H(X|Y)$, is defined as

$$H(X|Y) := - \sum_{xy \in \mathscr{X} \times \mathscr{Y}} P_{XY}(xy) \log P_{XY}(x|y) . \qquad (14)$$

The **entropy rate** of a stochastic process $\{X_t\}_{t \in T}$ is defined as

$$h = \lim_{n \to \infty} \frac{1}{n} H(X_1 X_2 \ldots X_n) . \qquad (15)$$

A different definition of entropy rate is as follows:

$$h' = \lim_{n \to \infty} H(X_n | X_1 \ldots X_{n-1}) . \qquad (16)$$

For stationary stochastic processes, $h = h'$. The entropy rate $H(X_n | X_1 \ldots X_{n-1})$, for finite n, is denoted by h_n.

Shannon introduced the **mutual information** as a measure of correlation between two random variables X and Y, defined as follows:

$$I(X;Y) := \sum_{xy \in \mathscr{X} \times \mathscr{Y}} P_{XY}(xy) \log \frac{P_{XY}(xy)}{P_X(x) P_Y(y)} . \qquad (17)$$

The mutual information is a measure of the predictability of one random variable when the outcome of the other is known. Note that the mutual information is symmetric in its arguments and hence measures the amount of information 'shared' by the two variables. The mutual information is a general correlation function for two random variables, measuring both linear and non-linear correlations. In contrast to the covariance and many other correlation measures, it is also applicable to non-numeric random variables such as the distribution of words in an English text or the distribution of amino acids in a DNA sequence. This is one reason why it is widely used in complex systems research.

C   Algorithmic Information Theory

A mathematical formalisation of randomness and information without reference to probabilities was developed independently by the Soviet mathematician Andrey Kolmogorov and the American mathematicians Ray J. Solomonoff and Gregory Chaitin in the 1960s. They considered information as a property of a single message, rather than of a set of messages and their probabilities. A message is a string of letters from an alphabet, such as the Roman alphabet or the binary characters 0 and 1. An example of a string is 'Hello, World!'. The string is composed of letters from the Roman alphabet and from a set containing the comma and space characters and the exclamation mark.

The **algorithmic information** content of a string is, roughly speaking, the length of the shortest computer program which outputs the string. For the string 'Hello, World!', this is probably a program of the form 'print("Hello, World!")' which has roughly the same length as the string itself. However, for a string of 10,000 zeros and ones alternating, the shortest program is shorter than the string itself, and the string is called 'compressible'. The notion of compressibility is meaningful only with longer strings. Only perfectly random strings are completely incompressible, therefore algorithmic information is a measure of randomness. It can be confusing that the term 'information' is used for randomness, but one may think of randomness as the amount of information which has to be communicated to reproduce the string 'exactly', irrespective of how interesting the string is in other respects.

The precise definition of algorithmic information is as follows (Li and Vitányi 2009). Consider a string x, a computing device $\mathcal{U}$ and programs p of length $l(p)$. The algorithmic information of the string, $K(x)$, is the length of the shortest program p, which, when fed into a machine $\mathcal{U}$, produces output x, $\mathcal{U}(p) = x$,

$$K_{\mathcal{U}}(x) = \min_{p:\mathcal{U}(p)=x} l(p) . \tag{18}$$

The minimisation is done over all possible programs. There is a fundamental problem with carrying out the minimisation procedure: whether an arbitrary program will finish or run forever cannot be known in general. This is called the halting problem. It is one of the deepest results in computer science, and due to the British mathematician Alan Turing. As a consequence, the algorithmic information is not computable in principle, though it can often be approximated in practice. Other names for algorithmic information are 'algorithmic complexity' or 'Kolmogorov complexity'.

The fundamental insight of Kolmogorov, Solomonoff and Chaitin is that the minimum length of a program is independent of the computing device on which it is run (up to some constant which is independent of the string). Hence, the definition of algorithmic information refers to a universal computer, which is a fundamental notion introduced by Alan Turing in the 1940s. Algorithmic information is therefore a 'universal' notion of randomness for strings because it is context- (machine-) independent. On the other hand, the Shannon entropy is context-dependent, since it may assign different amounts of information to the same string when it is embedded in different sets with different probabilities.

The length is not the only important parameter of a program; its running time is of equal importance. There are very short programs that take a long time to run, while the print program might be long but finished very quickly. This trade-off is relevant to the measures of complexity (discussed at the end of Chapter 4), which include the well-known logical depth.

D Complex Networks

A **network**, or a graph, is a set of nodes and, for simplicity here, there is at most one edge between any ordered pair of nodes. Nodes and edges are also called vertices and links, respectively. In a **directed network** each edge has a directionality, beginning at one node and ending at another. In an undirected network there is no such distinction between the start and end node of an edge. An example of an undirected network is the Internet. The servers are nodes, and edges between them are the physical wirings. An example of a directed network is an ecological food web. Two animals are linked if one of them feeds on the other so that a predator has a directed edge to its prey.

The **degree** of a node in a network is the number of edges attached to it. In a directed network, one distinguishes between in-degree and out-degree. The in-degree of a node is the number of edges directed to the node, and the out-degree is the number of nodes directed away from it.

Mathematically, a network of n nodes is represented by an **adjacency matrix**, A, which is an $n \times n$ matrix where each non-zero entry A_{ij} represents an edge from node i to node j (Newman 2010). In an unweighted network the A_{ij} are 1 if an edge exists from node i to node j and 0 otherwise. A **weighted network** assigns a real number to each edge, $A_{ij} \in \mathbb{R}$. Such weights could, for example, represent the volume of data traffic between two servers. In an undirected network, $A_{ij} = A_{ji}$ since A_{ij} and A_{ji} refer to the same object. The in-degree of a node i is the number of non-zero entries in the jth column of

A. The out-degree of a node i is the number of non-zero entries in the ith row of A. In an undirected network, these numbers are equal.

The **degree distribution** of a network is the frequency distribution over node degrees. A uniform degree distribution, for example, means that nodes of degree 1 are equally likely as nodes of degree n. A **path** is a sequence of nodes such that every two consecutive nodes in the sequence are connected by an edge. In a directed network the nodes have to be connected by edges that all point in the forward direction. The **path length** is the number of edges traversed along the sequence of a path. The **shortest path** between two nodes is the sequence with the minimum number of traversed edges to get from one node to the other. The average shortest path is the sum of all shortest path lengths divided by their number. The **diameter** of a network is the longest of all shortest paths.

Bibliography

L. A. Adamic and B. A. Huberman. Power-law distribution of the world wide web. *Science*, 287(5461):2115, 2000.

R. Albert and A.-L. Barabási. Statistical mechanics of complex networks. *Reviews of Modern Physics*, 74(1):47–97, 2002.

R. Albert, H. Jeong, and A. L. Barabási. Diameter of the world-wide web. *Nature*, 401(6749):130, 1999.

R. Albert, H. Jeong, and A.-L. Barabási. Error and attack tolerance of complex networks. *Nature*, 406(6794):378–382, 2000.

Michel Anctil. *Dawn of the neuron: The early struggles to trace the origin of nervous systems*. McGill-Queen's Press, 2015.

P. W. Anderson. More is different. *Science*, 177(4047):393–396, 1972.

W. B. Arthur. Complexity and the economy. *Science*, 284(5411):107–109, 1999.

J.-P. Aubin. A survey of viability theory. *SIAM Journal on Control and Optimization*, 28(4):749–788, 1990.

Jean-Pierre Aubin. *Viability Theory*. Springer Science & Business Media, 2009.

C. Aymanns, J. D. Farmer, A. M. Kleinnijenhuis, and T. Wetzer. Models of financial stability and their application in stress tests. *Handbook of Computational Economics*, 4:329–391, 2018.

F. A. C. Azevedo et al. Equal numbers of neuronal and nonneuronal cells make the human brain an isometrically scaled-up primate brain. *Journal of Comparative Neurology*, 513(5):532–541, 2009.

P. Bak, C. Tang, and K. Wiesenfeld. Self-organized criticality. *Physical Review A*, 38(1):364, 1988.

Per Bak. *How Nature Works: The Science of Self-Organised Criticality*. Copernicus Press, 1996.

A.-L. Barabási and R. Albert. Emergence of scaling in random networks. *Science*, 286(5439):509–512, 1999.

M. Baringer and J. C. Larsen. Sixteen years of Florida current transport at 27°N. *Geophysical Research Letters*, 28(16):3179–3182, 2001.

S. Battiston et al. Complexity theory and financial regulation. *Science*, 351 (6275):818–819, 2016.

Eric D. Beinhocker. *The Origin of Wealth: Evolution, Complexity, and the Radical Remaking of economics*. Harvard Business Press, 2006.

A. Bekker et al. Dating the rise of atmospheric oxygen. *Nature*, 427(6970): 117–120, 2004.

C. Béné and L. Doyen. From resistance to transformation: A generic metric of resilience through viability. *Earth's Future*, 6(7):979–996, 2018.

L. Benedetti-Cecchi, L. Tamburello, E. Maggi, and F. Bulleri. Experimental perturbations modify the performance of early warning indicators of regime shift. *Current Biology*, 25(14):1867–1872, 2015.

C. H. Bennett. Logical depth and physical complexity. In Rolf Herken, editor, *The Universal Turing Machine – a Half-Century Survey*, pages 227–257. Oxford University Press, 1991.

T. Berners-Lee, R. Cailliau, J.-F. Groff, and B. Pollermann. World-wide web: The information universe. *Internet Research*, 20(4):461–471, 2010.

Ludwig von Bertalanffy. *General System Theory: Foundations, Development, Applications*. George Braziller, 1969.

L. M. A. Bettencourt. The origins of scaling in cities. *Science*, 340(6139): 1438–1441, 2013.

L. M. A. Bettencourt, J. Lobo, D. Helbing, C. Kühnert, and G. B. West. Growth, innovation, scaling, and the pace of life in cities. *Proceedings of the National Academy of Sciences*, 104(17):7301–7306, 2007.

W. Bialek, I. Nemenman, and N. Tishby. Predictability, complexity, and learning. *Neural Computation*, 13(11):2409–2463, 2001.

W. Bialek et al. Statistical mechanics for natural flocks of birds. *Proceedings of the National Academy of Sciences*, 109(13):4786–4791, 2012.

James Binney. *The Theory of critical phenomena : an introduction to the renormalization group*. Clarendon Press, 1992.

H. M. Blalock, editor. *Causal Models in the Social Sciences*. Routledge, 1985.

U. Brandes et al. On modularity clustering. *IEEE Transactions on Knowledge and Data Engineering*, 20(2):172–188, 2007.

K. H. Britten, M. N. Shadlen, W. T. Newsome, and J. A. Movshon. Responses of neurons in macaque MT to stochastic motion signals. *Visual Neuroscience*, 10(6):1157–1169, 1993.

W. Buffett. Annual report, 2002.
http://www.berkshirehathaway.com/2002ar/2002ar.pdf.

E. Bullmore and O. Sporns. Complex brain networks: Graph theoretical analysis of structural and functional systems. *Nature Reviews Neuroscience*, 10(3):186–198, 2009.

E. Bullmore and O. Sporns. The economy of brain network organization. *Nature Reviews Neuroscience*, 13(5):336, 2012.

C. Burr, N. Cristianini, and J. Ladyman. An analysis of the interaction between intelligent software agents and human users. *Minds and Machines*, 28(4):735–774, 2018.

J. Butterfield. Emergence, reduction and supervenience: A varied landscape. *Foundations of Physics*, 41(6):920–959, 2011a.

J. Butterfield. Less is different: Emergence and reduction reconciled. *Foundations of Physics*, 41(6):1065–1135, 2011b.

S. R. Carpenter et al. Early warnings of regime shifts: A whole-ecosystem experiment. *Science*, 332(6033):1079–1082, 2011.

E. Chaisson. Using complexity science to search for unity in the natural sciences. In C. H. Lineweaver, editor, *Complexity and the Arrow of Time*. Cambridge University Press, 2013.

Austin Chambers. *Modern Vacuum Physics*. CRC Press, 2004.

A. Clark. Whatever next? predictive brains, situated agents, and the future of cognitive science. *Behavioral and Brain Sciences*, 36(3):181–204, 2013.

P. U. Clark, N. G. Pisias, T. F. Stocker, and A. J. Weaver. The role of the thermohaline circulation in abrupt climate change. *Nature*, 415(6874): 863–869, 2002.

A. Clauset, C. Moore, and M. Newman. Hierarchical structure and the pre-diction of missing links in networks. *Nature*, 453(7191):98–101, 2008.

A. Clauset, C. R. Shalizi, and M. Newman. Power-law distributions in em-pirical data. *SIAM Review*, 51(4):661–703, 2009.

Computational Mechanics Group, 2015. http://cmpy.csc.ucdavis.edu/.

G. Coricelli et al. Regret and its avoidance: A neuroimaging study of choice behavior. *Nature Neuroscience*, 8(9), 2005.

I. D. Couzin and N. R. Franks. Self-organized lane formation and optimized traffic flow in army ants. *Proceedings of the Royal Society of London B: Biological Sciences*, 270(1511):139–146, 2003.

I. D. Couzin and J. Krause. Collective memory and spatial sorting in animal groups. *Theoretical Biology*, 218, 2002.

Thomas M. Cover and Joy A. Thomas. *Elements of Information Theory*. Wiley-Blackwell, 2nd edition, 2006.

Credit Suisse Research Institute. Global wealth databook. Technical report, 2016.

J. P. Crutchfield. The calculi of emergence: Computation, dynamics and induction. *Physica D: Nonlinear Phenomena*, 75(1-3):11–54, 1994.

J. P. Crutchfield. Between order and chaos. *Nature Physics*, 8(1):17, 2012.

J. P. Crutchfield and D. P. Feldman. Regularities unseen, randomness ob-served: Levels of entropy convergence. *Chaos: An Interdisciplinary Jour-nal of Nonlinear Science*, 13(1):25–54, 2003.

J. P. Crutchfield and C. R. Shalizi. Thermodynamic depth of causal states: Objective complexity via minimal representations. *Physical Review E*, 59 (1):275, 1999.

J. P. Crutchfield and K. Young. Inferring statistical complexity. *Physical Review Letters*, 63(2):105, 1989.

J. P Crutchfield, C. J. Ellison, and J. R. Mahoney. Time's barbed arrow: Irreversibility, crypticity, and stored information. *Physical Review Letters*, 103(9):094101, 2009.

V. Dakos et al. Slowing down as an early warning signal for abrupt climate change. *Proceedings of the National Academy of Sciences*, 105(38): 14308–14312, 2008.

A. Davidson. How AIG fell apart. Technical report, Reuter, 2008. http://www.reuters.com/article/us-how-aig-fell-apart-idUSMAR85972720080918.

P. Davies and N. H. Gregersen, editors. *Information and the Nature of Reality: From Physics to Metaphysics*. Cambridge University Press, 2014.

Sybren Ruurds De Groot and Peter Mazur. *Non-Equilibrium Thermodynamics*. Courier Corporation, 2013.

H. de Jong. Modeling and simulation of genetic regulatory systems: A literature review. *Journal of Computational Biology*, 9(1):67–103, 2002.

J. DeFelipe, L. Alonso-Nanclares, and J. I. Arellano. Microstructure of the neocortex: Comparative aspects. *Journal of neurocytology*, 31(3-5):299–316, 2002.

Guillaume Deffuant and Nigel Gilbert. *Viability and Resilience of Complex Systems: Concepts, Methods and Case Studies from Ecology and Society*. Springer Science & Business Media, 2011.

A. J. Dunleavy, K. Wiesner, R. Yamamoto, and C. P. Royall. Mutual information reveals multiple structural relaxation mechanisms in a model glass former. *Nature communications*, 6:6089, 2015.

J. A. Dunne, R. J. Williams, and N. D. Martinez. Food-web structure and network theory: The role of connectance and size. *Proceedings of the National Academy of Sciences*, 99(20):12917–12922, 2002.

David Easley and Jon Kleinberg. *Networks, Crowds, and Markets: Reasoning About a Highly Connected World*. Cambridge University Press, 2010.

Joshua M. Epstein and Robert Axtell. *Growing Artificial Societies: Social Science from the Bottom Up*. Brookings Institution Press and MIT Press, 1996.

P. Erdös and A. Rényi. On the evolution of random graphs. *Publication of the Mathematical Institute of the Hungarian Academy of Sciences*, 5(1): 17–60, 1960.

A. Fabiani, F. Galimberti, S. Sanvito, and A. R. Hoelzel. Extreme polygyny among southern elephant seals on sea lion island, falkland islands. *Behavioral Ecology*, 15(6):961–969, 2004.

Kenneth Falconer. *Fractal Geometry: Mathematical Foundations and Applications*. John Wiley & Sons, 2004.

P. G. Falkowski, T. Fenchel, and E. F. Delong. The microbial engines that drive earth's biogeochemical cycles. *Science*, 320(5879):1034–1039, 2008.

M. Faloutsos, P. Faloutsos, and C. Faloutsos. On power-law relationships of the internet topology. *ACM SIGCOMM Computer Communication Review*, 29(4):251–262, 1999.

E. F. Fama. Efficient capital markets: II. *The Journal of Finance*, 46(5): 1575–1617, 1991.

J. D. Farmer. Physicists attempt to scale the ivory towers of finance. *Computing in Science & Engineering*, 1(6):26–39, 1999.

J. D. Farmer and J. Geanakoplos. The virtues and vices of equilibrium and the future of financial economics. *Complexity*, 14(3):11–38, 2009.

J. D. Farmer and N. Packard, editors. *Evolution, Games, and Learning: Models for Adaptation in Machines and Nature – Proceedings of the Fifth Annual International Conference, May 20-24, 1985*, volume 22, 1-3 1986. Physica D: Nonlinear Phenomena.

Richard P. Feynman. *Feynman Lectures on Computation*. Addison-Wesley, 1998.

Stanley Finger. *Minds Behind the Brain: A History of the Pioneers and Their Discoveries*. Oxford University Press, 2005.

R. A. Fisher. The wave of advance of advantageous genes. *Annals of Eugenics*, 7(4):355–369, 1937.

R. Foote. Mathematics and complex systems. *Science*, 318(5849):410–412, 2007.

S. Fortunato and M. Barthelemy. Resolution limit in community detection. *Proceedings of the National Academy of Sciences*, 104(1):36–41, 2007.

N. R Franks. Army ants: a collective intelligence. *American Scientist*, 77: 138–145, 1989.

N. R. Franks, N. Gomez, S. Goss, and J. L. Deneubourg. The blind leading the blind in army ant raid patterns: Testing a model of self-organization (Hymenoptera: Formicidae). *Journal of Insect Behavior*, 4(5):583–607, 1991.

K. Friston. Causal modelling and brain connectivity in functional magnetic resonance imaging. *PLoS Biology*, 7(2):e1000033, 2009.

S. Garnier et al. Stability and responsiveness in a self-organized living architecture. *PLoS Computational Biology*, 9(3):e1002984, 2013.

M. Gell-Mann. What is complexity? *Complexity*, 1(1):16–19, 1995.

M. Gell-Mann and S. Lloyd. Information measures, effective complexity, and total information. *Complexity*, 2(1):44–52, 1996.

Murray Gell-Mann. *The Quark and the Jaguar*. W. H. Freeman, 1994.

N. Goldenfeld and L. P. Kadanoff. Simple lessons from complexity. *Science*, 284(5411):87–89, 1999.

Herbert Goldstein. *Classical Mechanics*. Addison-Wesley series in advanced physics. Addison-Wesley, 1950.

Deborah M. Gordon. *Ant Encounters: Interaction Networks and Colony Behavior*. Princeton University Press, 2010.

Stephen Jay Gould. *Full House*. Harvard University Press, 2011.

P. Grassberger. Toward a quantitative theory of self-generated complexity. *International Journal of Theoretical Physics*, 25(9):907–938, 1986.

Richard K. Guy and John H. Conway. *Winning Ways for Your Mathematical Plays*. Academic Press, 1982.

A. G. Haldane. Rethinking the financial network, 2009. https://www.bankofengland.co.uk/speech/2009/rethinking-the-financial-network.

A. G. Haldane and R. M. May. Systemic risk in banking ecosystems. *Nature*, 469(7330):351–355, 2011.

Heiko Hamann. *Swarm Robotics: A Formal Approach*. Springer, 2018.

D. Haymann. How SARS was contained, 2013. http://www.nytimes.com/2013/03/15/opinion/global/how-sars-was-contained.html.

R. M. Hazen, D. Papineau, and W. Bleeker. Mineral evolution. *American Mineralogist*, 93:1693–1720, 2008.

A. D. Henry, P. Prałat, and C.-Q. Zhang. Emergence of segregation in evolving social networks. *Proceedings of the National Academy of Sciences*, 108(21):8605–8610, 2011.

J. H. Holland. Complex adaptive systems. *Daedalus*, 121(1):17–30, 1992.

John H. Holland. *Complexity: A Very Short Introduction*. Oxford University Press, 2014.

Bert Hölldobler and Edward O. Wilson. *The Superorganism : the Beauty, Elegance, and Strangeness of Insect Societies*. W.W. Norton, 2008.

C. S. Holling. Resilience and stability of ecological systems. *Annual Review of Ecology and Systematics*, 4(1):1–23, 1973.

B. Holmstrom and J. Tirole. Financial intermediation, loanable funds, and the real sector. *The Quarterly Journal of Economics*, 112(3):663–691, 1997.

John E. Hopcroft, Rajeev Motwani, and Jeffrey D. Ullman. *Introduction to Automata Theory, Languages, and Computation*. Addison-Wesley, 2nd edition, 2001.

S. Houman. Credit default swaps and regulatory reform. Technical report, Mercatus Center, George Mason University, 2009. https://www.mercatus.org/publication/credit-default-swaps-and-regulatory-reform.

B. A. Huberman and L. A. Adamic. Evolutionary dynamics of the world wide web. *arXiv preprint cond-mat/9901071*, 1999.

Paul Humphreys. *Emergence: A Philosophical Account*. Oxford University Press, 2016.

IEEE Communications Society. Infographic: The internet of things, 2015. https://www.comsoc.org/blog/infographic-internet-things-iot.

Nobuyuki Ikeda and Shinzo Watanabe. *Stochastic Differential Equations and Diffusion Processes*. North-Holland, 2014.

International Telecommunication Union. Overview of the internet of things. Technical report, 2012. https://www.itu.int/rec/T-REC-Y.2060-201206-I.

Internet Systems Consortium. Internet host count history. Technical report, 2012. https://www.isc.org/network/survey/, Accessed August 2018.

IPCC 2013. Technical summary. In T. F. Stocker et al., editors, *Climate Change 2013: The Physical Science Basis. Contribution of Working Group I to the Fifth Assessment Report*. Cambridge University Press, 2013a.

IPCC 2013. Carbon and other biogeochemical cycles. In T. F. Stocker et al., editors, *Climate Change 2013: The Physical Science Basis. Contribution of Working Group I to the Fifth Assessment Report of the Intergovernmental Panel on Climate Change*. Cambridge University Press, 2013b.

IPCC 2013. Introduction. In T.F. Stocker et al., editors, *Climate Change 2013: The Physical Science Basis. Contribution of Working Group I to the Fifth Assessment Report of the Intergovernmental Panel on Climate Change*. Cambridge University Press, 2013c.

D. Jardim-Messeder et al. Dogs have the most neurons, though not the largest brain: Trade-off between body mass and number of neurons in the cerebral cortex of large carnivoran species. *Frontiers in Neuroanatomy*, 11:118, 2017.

E. T. Jaynes. Information theory and statistical mechanics. *Physical Review*, 106(4):620, 1957.

H. Jeong, S. P. Mason, A.-L. Barabási, and Z. N. Oltvai. Lethality and centrality in protein networks. *Nature*, 411(6833):41, 2001.

H. Jeong et al. The large-scale organization of metabolic networks. *Nature*, 407(6804):651, 2000.

L. Jost. Entropy and diversity. *Oikos*, 113(2):363–375, 2006.

D. Kahneman. A psychological perspective on economics. *American economic review*, 93(2):162–168, 2003.

J. D. Keeler and J. D. Farmer. Robust space-time intermittency and 1f noise. *Physica D: Nonlinear Phenomena*, 23(1–3):413–435, 1986.

D. Kelly, M. Dillingham, A. Hudson, and K. Wiesner. A new method for inferring hidden markov models from noisy time sequences. *PloS One*, 7 (1):e29703, 2012.
http://www.mathworks.com/matlabcentral/fileexchange/33217.

S.-H. Kim. Fractal structure of a white cauliflower. *arXiv preprint cond-mat/0409763*, 2004.

M. Kleiber. Body size and metabolism. *Hilgardia*, 6(11):315–353, 1932.

A. Klein et al. Evolution of social insect polyphenism facilitated by the sex differentiation cascade. *PLoS Genetics*, 12(3):e1005952, 2016.

D. Krakauer. The inferential evolution of biological complexity: Forgetting nature by learning to nurture. In C. H. Lineweaver, editor, *Complexity and the Arrow of Time*. Cambridge University Press, 2013.

J. Ladyman, J. Lambert, and K. Wiesner. What is a complex system? *European Journal for Philosophy of Science*, 3(1):33–67, 2013.

S. Lawrence and C. L. Giles. Accessibility of information on the web. *Nature*, 400(6740):107–109, 1999.

C.-B. Li, H. Yang, and T. Komatsuzaki. Multiscale complex network of protein conformational fluctuations in single-molecule time series. *Proceedings of the National Academy of Sciences*, 105(2):536–541, 2008.

Ming Li and Paul Vitányi. *An Introduction to Kolmogorov Complexity and Its Applications*. Springer, 3rd edition, 2009.

C. Lineweaver. A simple treatment of complexity: Cosmological entropic boundary conditions on increasing complexity. In C. H. Lineweaver, editor, *Complexity and the Arrow of Time*. Cambridge University Press, 2013.

S. Lloyd. Measures of complexity: A nonexhaustive list. *Control Systems Magazine, IEEE*, 21(4):7–8, 2001.

S. Lloyd and H. Pagels. Complexity as thermodynamic depth. *Annals of Physics*, 188(1):186–213, 1988.

Seth Lloyd. *Programming the Universe*. Knopf, 2006.

E. Lorenz. Predictability: Does the flap of a butterfly's wings in brazil set off a tornado in texas? Presentation at American Association for the Advancement of Science, 139th meeting, Washington, DC, 1972. http://eaps4.mit.edu/research/Lorenz/Butterfly_1972.pdf.

J. E. Lovelock and L. Margulis. Atmospheric homeostasis by and for the biosphere: The gaia hypothesis. *Tellus*, 26(1-2):2–10, 1974.

R. S. MacKay. Nonlinearity in complexity science. *Nonlinearity*, 21:T273, 2008.

B. F. Madore and W. L. Freedman. Computer simulations of the Belousov-Zhabotinsky reaction. *Science*, 222(4624):615–616, 1983.

Klaus Mainzer. *Thinking in Complexity – The Computational Dynamics of Matter*. Springer, 1st edition, 1994.

E. B. Mallon and N. R. Franks. Ants estimate area using Buffon's needle. *Proceedings of the Royal Society of London B: Biological Sciences*, 267 (1445):765–770, 2000.

Benoît B. Mandelbrot. *The Fractal Geometry of Nature*. W. H. Freeman, 1983.

Benoît B. Mandelbrot. *Fractals and Scaling in Finance: Discontinuity, Concentration, Risk*. Springer Science & Business Media, 2013.

Benoît B. Mandelbrot and Richard Hudson. *The Mis Behaviour of Markets: A Fractal View of Risk, Ruin and Reward*. Profile Books, 2010.

M. A. Marra et al. The genome sequence of the sars-associated coronavirus. *Science*, 300(5624):1399–1404, 2003.

Alfred Marshall. *Principles of Economics*. Macmillan, 1890.

Mark Maslin. *Climate: A Very Short Introduction*. Oxford University Press, 2013.

H. R. Mattila and T. D. Seeley. Genetic diversity in honey bee colonies enhances productivity and fitness. *Science*, 317(5836):362–364, 2007.

J. W. McAllister. Effective complexity as a measure of information content. *Philosophy of Science*, 70(2):302–307, 2003.

Hans Meinhardt. *Models of Biological Pattern Formation*. Academic Press, 1982.

R. Menzel and M. Giurfa. Cognitive architecture of a mini-brain: The honeybee. *Trends in Cognitive Sciences*, 5(2):62–71, 2001.

S. Milgram. The small world problem. *Psychology today*, 2(1):60–67, 1967.

Melanie Mitchell. *Complexity: A Guided Tour*. Oxford University Press, 2011.

M. Mitzenmacher. A brief history of generative models for power law and lognormal distributions. *Internet Mathematics*, 1(2):226–251, 2004.

J. M. Montoya and R. V. Solé. Small world patterns in food webs. *Journal of Theoretical Biology*, 214(3):405–412, 2002.

Moz. https://moz.com/top500/pages, 2018. accessed on 24 August 2018.

Carl D. Murray and Stanley F. Dermott. *Solar System Dynamics*. Cambridge University Press, 1999.

M. Newman. The structure and function of complex networks. *SIAM Review*, 45(2), 2003.

M. Newman. Power laws, Pareto distributions and Zipf's law. *Contemporary Physics*, 46(5):323–351, 2005.

M. Newman and M. Girvan. Finding and evaluating community structure in networks. *Physical review E*, 69(2):026113, 2004.

M. Newman, S. Forrest, and J. Balthrop. Email networks and the spread of computer viruses. *Physical Review E*, 66(3):035101, 2002.

Mark Newman. *Networks: An Introduction*. Oxford University Press, 1st edition, 2010.

Alyssa Ney. *Metaphysics: An Introduction*. Routledge, 2014.

Grégoire Nicolis and Iiya Prigogine. *Self-Organzisation in Nonequilibrium Systems*. John Wiley & Sons, 1977.

John Nolte and John W. Sundsten. *The Human Brain : an Introduction to its Functional Anatomy*. Mosby, 2002.

A. Nordrum. Popular internet of things forecast of 50 billion devices by 2020 is outdated. Technical report, IEEE Spectrum, 2016. http://spectrum.ieee.org/tech-talk/telecom/internet/popular-internet-of-things-forecast-of-50-billion-devices-by-2020-is-outdated.

P. Nurse. Life, logic and information. *Nature*, 454(7203):424–426, 2008.

R. K. Pachauri, L. Meyer, and Core Writing Team. Climate change 2014: Synthesis report. contribution of working groups I, II and III to the fifth assessment report of the Intergovernmental Panel on Climate Change. Technical report, 2014.

Scott E. Page. *Diversity and Complexity*. Princeton University Press, 1st edition, 2010.

A. J. Palmer, C. W. Fairall, and W. A. Brewer. Complexity in the atmosphere. *IEEE Transactions on Geoscience and Remote Sensing*, 38(4):2056–2063, 2000.

S. E. Palmer, O. Marre, M. J. Berry, and W. Bialek. Predictive information in a sensory population. *Proceedings of the National Academy of Sciences*, 112(22):6908–6913, 2015.

Vilfrido Pareto. *Manual of Political Economy*. Kelley, 1980. Reprint.

J. K. Parrish and L. Edelstein-Keshet. Complexity, pattern, and evolutionary trade-offs in animal aggregation. *Science*, 284(5411):99–101, 1999.

D. Paul. Credit default swaps, the collapse of AIG and addressing the crisis of confidence. *Huffington Post*, 2008. http://www.huffingtonpost.com/david-paul/credit-default-swaps-the_b_133891.html.

Azaria Paz. *Introduction to Probabilistic Automata*. Academic Press, 1971.

Roger Penrose. *The Road to Reality*. Jonathan Cape, 2004.

Thomas Piketty. *Capital in the Twenty-First Century*. Harvard University Press, 2014.

Stuart L. Pimm. *Food Webs*. Springer, 1982.

D. Pines, editor. *Emerging Syntheses in Science: Proceedings from the Founding Workshops of the Santa Fe Institute*. SFI Press, 2019.

D. Price. A general theory of bibliometric and other cumulative advantage processes. *Journal of the Association for Information Science and Technology*, 27(5):292–306, 1976.

I. Prigogine. Time, structure, and fluctuations. *Science*, 201(4358):777–785, 1978.

Ilya Prigogine. *From Being to Becoming Time and Complexity in the Physical Sciences*. W. H. Freeman, 1980.

Gunnar Pruessner. *Self-Organised Criticality: Theory, Models and Characterisation*. Cambridge University Press, 2012.

D. Purves et al., editors. *Neuroscience*. Oxford University Press, 2018.

S. Rahmstorf. The thermohaline ocean circulation: A system with dangerous thresholds? *Climatic Change*, 46(3):247–256, 2000.

K. Rauss, S. Schwartz, and G. Pourtois. Top-down effects on early visual processing in humans: A predictive coding framework. *Neuroscience & Biobehavioral Reviews*, 35(5):1237–1253, 2011.

E. Ravasz and A.-L. Barabási. Hierarchical organization in complex networks. *Physical Review E*, 67(2):026112, 2003.

C. R. Reid et al. Army ants dynamically adjust living bridges in response to a cost-benefit trade-off. *Proceedings of the National Academy of Sciences*, 112(49):15113–15118, 2015.

D. Rind. Complexity and climate. *Science*, 284(5411):105–107, 1999.

Gerald H. Ristow. *Pattern formation in granular materials*. Springer, 2000.

James William Rohlf. *Modern Physics from α to Z^0*. Wiley, 1994.

N. Rooney, K. McCann, G. Gellner, and J. C. Moore. Structural asymmetry and the stability of diverse food webs. *Nature*, 442(7100):265, 2006.

A. Rosenblueth, N. Wiener, and J. Bigelow. Behavior, purpose and teleology. *Philosophy of Science*, 10(1):18–24, 1943.

Don Ross. *Philosophy of Economics*. Palgrave Macmillan, 2014.

M. Rosvall and C. T. Bergstrom. Maps of random walks on complex networks reveal community structure. *Proceedings of the National Academy of Sciences*, 105(4):1118–1123, 2008.

G. Sargut and R. G. McGrath. Learning to live with complexity. *Harvard Business Review*, 89(9):68–76, 2011.

M. Scheffer. Complex systems: Foreseeing tipping points. *Nature*, 467 (7314):411–412, 2010.

M. Scheffer, S. R Carpenter, V. Dakos, and E. H. van Nes. Generic indicators of ecological resilience: Inferring the chance of a critical transition. *Annual Review of Ecology, Evolution, and Systematics*, 46:145–167, 2015.

T. C. Schelling. Models of segregation. *The American Economic Review*, 59 (2):488–493, 1969.

E. Schneidman, M. J. Berry, R. Segev, and W. Bialek. Weak pairwise correlations imply strongly correlated network states in a neural population. *Nature*, 440(7087):1007–1012, 2006.

T. Schwander et al. Nature versus nurture in social insect caste differentiation. *Trends in ecology & evolution*, 25(5):275–282, 2010.

Thomas D. Seeley. *The Wisdom of the Hive: The Social Physiology of Honey Bee Colonies*. Harvard University Press, 2009.

Thomas D. Seeley. *Honeybee Democracy*. Princeton University Press, 2010.

C. R. Shalizi and J. P. Crutchfield. Computational mechanics: Pattern and prediction, structure and simplicity. *Journal of Statistical Physics*, 104(3): 817–879, 2001.

C. R. Shalizi and K. Klinkner. An algorithm for building markov models from time series, 2003. http://bactra.org/CSSR/.

C. R. Shalizi, K. L. Shalizi, and R. Haslinger. Quantifying self-organization with optimal predictors. *Physical Review Letters*, 93(11):118701, 2004.

C. E. Shannon. A mathematical theory of communication. Technical report, Bell Labs, 1948.

H. A. Simon. The architecture of complexity. *Proceedings of the American Philosophical Society*, 106(6):467–482, 1962.

H. A. Simon. How complex are complex systems? *Proceedings of the Biennial Meeting of the Philosophy of Science Association*, 2:507–522, 1976.

Adam Smith. *An Inquiry into the Nature and Causes of the Wealth of Nations*. Strahan, 1776.

E. Smith. Emergent order in processes: The interplay of complexity, robustness, correlation, and hierarchy in the biosphere. In C. H. Lineweaver, editor, *Complexity and the Arrow of Time*. Cambridge University Press, 2013.

Eric Smith and Harold J Morowitz. *The Origin and Nature of Life on Earth: The Emergence of the Fourth Geosphere*. Cambridge University Press, 2016.

Kim Sneppen and Giovanni Zocchi. *Physics in Molecular Biology*. Cambridge University Press, 2005.

D. Sornette. Predictability of catastrophic events: Material rupture, earthquakes, turbulence, financial crashes, and human birth. *Proceedings of the National Academy of Sciences*, 99:2522–2529, 2002.

Didier Sornette. *Why Stockmarkets Crash: Critical Events in Financial Markets*. Princeton University Press, 2003.

Didier Sornette. *Critical Phenomena in Natural Sciences: Chaos, Fractals, Selforganization and Disorder: Concepts and Tools*. Springer, 2nd edition, 2009.

G. J. Stigler. The development of utility theory. II. *Journal of Political Economy*, 58(5):373–396, 1950a.

G. J. Stigler. The development of utility theory. I. *Journal of Political Economy*, 58(4):307–327, 1950b.

D. B. Stouffer and J. Bascompte. Compartmentalization increases food-web persistence. *Proceedings of the National Academy of Sciences*, 108(9): 3648–3652, 2011.

M. Strevens. Complexity theory. In P. Humphreys, editor, *The Oxford Handbook of Philosophy of Science*. Oxford University Press, 2016.

Steven H. Strogatz. *Nonlinear Dynamics and Chaos: With Applications to Physics, biology, chemistry, and engineering*. Westview Press, 2014.

G. Sugihara and R. M. May. Applications of fractals in ecology. *Trends in Ecology & Evolution*, 5(3):79–86, 1990.

T. Sullivan. Embracing complexity. *Harvard Business Review*, 89(9):88–92, September 2011.

Nassim Nicholas Taleb. *The Black Swan: The Impact of the Highly Improbable*. Random House, 2007.

A. Toomre and J. Toomre. Galactic bridges and tails. *The Astrophysical Journal*, 178:623–666, 1972.

A. M. Turing. The chemical basis of morphogenesis. *Philosophical Transactions of the Royal Society of London. Series B, Biological Sciences*, 237 (641):37–72, 1952.

S. Ulam. Random processes and transformations. In *Proceedings of the International Congress on Mathematics*, volume 2, pages 264–275, 1952.

P. A. M. Van Dongen. *The Central Nervous System of Vertebrates*. Springer, 1998.

Nicolaas Godfried Van Kampen. *Stochastic Processes in Physics and Chemistry*, volume 1. Elsevier, 1992.

R. van Steveninck et al. Reproducibility and variability in neural spike trains. *Science*, 275(5307):1805–1808, 1997.

A. J. Veraart et al. Recovery rates reflect distance to a tipping point in a living system. *Nature*, 481(7381):357–359, 2012.

John von Neumann. *Theory of Self-Reproducing Automata*. University of Illinois Press, 1966. edited and completed by Arthur W. Burks.

John von Neumann and Oskar Morgenstern. *Theory of Games and Economic Behavior*. Princeton University Press, 1947.

Mitchell Waldrup. *Complexity: The Emerging New Paradigm at the Edge of Order and Chaos*. Prentice Hall, 1992.

N. S. Ward. Functional reorganization of the cerebral motor system after stroke. *Current Opinion in Neurology*, 17(6):725–730, 2004.

N. W. Watkins et al. 25 years of self-organized criticality: Concepts and controversies. *Space Science Reviews*, 198(1-4):3–44, 2016.

M. L. Weitzman. On diversity. *The Quarterly Journal of Economics*, 107(2): 363–405, 1992.

G. Weng, U. S. Bhalla, and R. Iyengar. Complexity in biological signaling systems. *Science*, 284(5411):92–96, 1999.

C. Werndl. What are the new implications of chaos for unpredictability? *The British Journal for the Philosophy of Science*, 60(1):195–220, 2009. ISSN 0007-0882.

B. T. Werner. Complexity in natural landform patterns. *Science*, 284(5411): 102–104, 1999.

G. B. West, J. H. Brown, and B. J. Enquist. A general model for the origin of allometric scaling laws in biology. *Science*, 276(5309):122–126, 1997.

J. G. White, E. Southgate, J. N. Thomson, and S. Brenner. The structure of the nervous system of the nematode Caenorhabditis elegans. *Philosophical Transactions of the Royal Society London B, Biological Sciences*, 314 (1165):1–340, 1986.

G. M. Whitesides and R. F. Ismagilov. Complexity in chemistry. *Science*, 284(5411):89–92, 1999.

Norbert Wiener. *Cybernetics: Or Control and Communication in the Animal and the Machine*. MIT Press, 2nd edition, 1961.

K. Wiesner, M. Gu, E. Rieper, and V. Vedral. Information-theoretic lower bound on energy cost of stochastic computation. *Proc. R. Soc. A*, 468 (2148):4058–4066, 2012.

J. Wilson. Metaphysical emergence: Weak and strong. In T. Bigaj and C. Wuthrich, editors, *Metaphysics in Contemporary Physics*, pages 251–306. Poznan Studies in the Philosophy of the Sciences and the Humanities, 2015.

S. Wolfram, editor. *Cellular Automata: Proceedings of an Interdisciplinary Workshop*, 1984. Center for Nonlinear Studies, Los Alamos, North-Holland.

Stephen Wolfram. *A New Kind of Science*. Wolfram Media, 2002.

D. Wolpert. Information width: A way for the second law to increase complexity. In C. H. Lineweaver, editor, *Complexity and the Arrow of Time*. Cambridge University Press, 2013.

World Health Organization. Severe acute respiratory syndrome (SARS): Status of the outbreak and lessons for the immediate future. Technical report, 2013. www.who.int/csr/media/sars_wha.pdf.

Worldwidewebsize. Daily estimated size of the World Wide Web. http://www.worldwidewebsize.com/, 2017. accessed 14 July 2017.

J. Z. Young. The number and sizes of nerve cells in octopus. *Proceedings of the Zoological Society of London*, 140(2):229–254, 1963.

J. Ziv and A. Lempel. A universal algorithm for sequential data compression. *IEEE Transactions on Information Theory*, 23(3):337–343, 1977.

G. Zöller, S. Hainzl, and J. Kurths. Observation of growing correlation length as an indicator for critical point behavior prior to large earthquakes. *Journal of Geophysical Research: Solid Earth*, 106(B2):2167–2175, 2001.

W. H. Zurek. Algorithmic randomness and physical entropy. *Physical Review A*, 40(8):4731–4751, 1989.

Index

Adaptive behaviour, 2, 4, 19, 37, 43, 59, 61, 65, 66, 70, 72, 73, 82, 83, 85, 123, 124, 126, 129, 130

Agent, 5, 6, 27, 28, 45, 47, 49–52, 67, 72, 73, 94, 119
  irrational, 48
  rational, 47, 48

Agent-based models, 49, 94

Albert, Reka, 106

Algorithm, 41, 112, 116, 128
  clustering, 107
  community detection, 108
  preferential attachment, 109

Algorithmic complexity, 110, 114–116, 135, 139

Algorithmic Information Theory, 139

Analogy between bee hives and brains, 44

Anderson, Philip, 3, 16, 22, 24, 81, 84

Ant colonies, 4, 38–41, 43, 81, 96, 126

Approximation, 5, 21–23, 116

Aristotle, 11

Attractors, 15, 103

Bak, Per, 123

Bank of England, 1, 52

Barabási, Albert-Lásló, 106

Beehives, 42–44, 69

compared to brains, 44

Beinhocker, Eric, 49

Belousov-Zhabotinsky (BZ) reaction, 4, 27, 70, 72–74, 77, 80, 128, 129

Bennett, Charles, 16, 80, 82, 115, 124

Bigelow, Julian, 12

Birds, 4, 6, 7, 67, 70, 71, 79, 82, 97, 125

Bouchaud, Jean-Philippe, 49

Boundaries, 23, 25, 32, 34, 64, 127

Brain, 6–8, 17, 26, 27, 44, 57–61, 66, 67, 69–71, 79, 81, 83, 85, 88, 91, 92, 98, 102, 105, 118, 119, 121, 124, 128, 129

Browder, Felix, 16

Buffon's Needle, 41

Butterfly effect, 14

BZ reaction, *see* Belousov-Zhabotinsky reaction.

C. Elegans, 57

Causal interactions, 6, 67, 81, 84, 101, 115, 116

CDO, *see* Collateralised debt obligation.

CDS, *see* Credit default swap.

Cellular automata, 12, 15, 16, 113

Chaos, 14, 15, 64, 78, 100, 122

Chemistry, 2, 11, 12, 16, 19, 21, 27, 74, 77, 102, 118, 120, 127

Chomsky hierarchy, 112
Climate system, 33–37, 103
Collateralised debt obligation (CDO), 51
Communication, 6, 12, 38, 56, 67, 79, 83, 137
Complex adaptive behaviour, 2, *see also* Adaptive behaviour.
Complex networks, *see* Networks.
Complex Systems Society, 17
Complexity
    conditions for, 10, 65, 66, 129, 130
    functional conceptions of, 124
    generic conceptions of, 121
    physical conceptions of, 122
    products of, 65
    truisms of, 9, 12, 21, 28, 65, 74, 75, 88, 118, 126, 127
Complicated, 3, 7, 22, 33, 34, 63, 64, 78, 101, 117, 126
Computation, 5, 7, 9, 12, 13, 15–17, 76, 82, 83, 85, 88, 95, 111–113, 119–122, 124–127, 137
Computational mechanics, 111–113
Conrad, Michael, 17
Consciousness, 8, 17, 37, 57, 60
Conway-Morris, Simon, 125
Correlation, 24, 26, 27, 67, 68, 71, 82, 89, 91, 92, 96–99, 101, 104, 108, 128, 138
    length, 104
    Pearson correlation coefficient, 97, 101
Correlation function, 96, 97, 104, 105, 109, 138
Coupled maps, 27
Covariance, 96, 97, 104, 136, 138
Cowan, George, 16, 17
Cowan, Jack, 16

Credit default swap (CDS), 51, 52
Critical phenomena, 3, 26–28, 67, 75, 127, 130, *see also* Self-organised criticality
Critical point, 26, 27, 50, 104, 105, 123, 127, 128
Critical slowing down, 103–105
Crutchfield, James, 16, 111, 113
Cybernetics, 12, 13

Descartes, René, 19, 20
Differential equations, 14, 22, 63, 93, 96, 100, 102
Disorder, 10, 65, 68–70, 73, 85, 89–92, 95, 111, 122, 130
Distribution
    degree distribution, 55, 56, 76, 91, 109, 141
    frequency distribution, 141
    Gaussian, 100
    of wealth, 46
    of words in English, 46, 138
    power-law, 49, 58, 100
    scale-free, 55, 56
    Zipf, 100
Diversity, 30, 52, 53, 58, 65, 68, 70, 73, 85, 89, 92, 93
    of type, 89, 92, 93
Dynamical systems theory, 8, 12–14, 16, 17, 100, 101, 103, 121
Dynamics, 7, 14, 15, 22, 23, 25, 27, 31–35, 49, 61, 70, 74, 75, 77–79, 89, 93, 94, 100–103, 110, 121, 123, 126–128

Economics, 2, 50
    behavioural economics, 48
    classical economics, 45
    financial economics, 52, 53
    macroeconomics, 50
    microeconomics, 47, 50

supply and demand curves in, 45
Economies, 124
    decision making economies, 44
    digital economies, 47
Econophysics, 50
Ecosystems, 52, 92, 103, 105, 124
    financial, 52
    marine, 105
Effective complexity, 87, 114, 115
Effective measure complexity, 87, 98
Eigen, Manfred, 16
Emergence, 3, 4, 15, 16, 21–24, 27,
        28, 37, 60, 65, 73, 74, 76,
        78, 79, 81, 84, 85, 121–123,
        127, 128
    complexity as, 121
    epistemological emergence, 121
    of dynamics, properties and laws,
        74
    of structure at different scales, 74
    of various kinds of invariance and
        forms of universal behaviour,
        75
    ontological emergence, 121
Energy, 21, 25, 27, 29, 36, 54, 59,
        67–69, 72, 73, 99
    free energy, 25, 71, 72, 123
Entropy, 33, 137
    entropy power, 92
    entropy rate, 91, 92, 111, 138
    excess entropy, 98, 113
    Shannon entropy, 90–92, 97, 113,
        114, 137, 140
    thermodynamic entropy, 71, 73,
        123
Equilibrium, 19, 27, 28, 46–50, 52,
        64, 73, 102, 123, 127
    chemical, 73, 96
    dynamic, 28, 70, 72, 73, 77
    market, 46

Nash, 28
    price, 45, 46
    stable, 45, 53
    static, 28
    thermodynamic (thermal), 28, 65,
        71–73, 121, 127
Erdös-Renyi random graph model, 90,
        91, 107
Error correction, 43, 71, 79, 80
Eusocial insects, 37, 43, 83
Expectation value, 136, 137

Farmer, Doyne, 16, 17, 49, 50
Feedback, 4, 7, 8, 10, 12, 14, 29–31,
        35, 36, 39, 40, 42, 43, 46,
        47, 59–61, 65, 70, 71, 73,
        79, 81, 83, 85, 93–95, 100,
        101, 109, 110, 127, 129, 130,
        133
    negative, 35, 37, 43–45, 47, 71,
        129
    positive, 35–38, 46, 71, 128, 129
Feldman, Marcus, 16
Fixed point, 102, 103
Flocking/flocks, 70, 71, 97
Food webs, 55, 102, 105, 107, 140
Forces, 20, 29, 32, 34, 67, 118
Fractal, 16, 24, 25, 108, 109
Frauenfelder, Hans, 16
Function, 10, 66
Function (purpose), 2, 10, 19, 37, 58,
        59, 63, 66, 71, 72, 79, 81–
        83, 85, 107, 124, 125, 127,
        128, 130, 131

Gaia hypothesis, 130
Galileo, 20
Game of Life, 15, 121
Game theory, 17, 28
Gassendi, Pierre, 19
Geanakoplos, John, 50

Gell-Mann, Murray, 2, 16, 87, 114, 115, 120
General system theory, 13
Girvan, Michelle, 108
Golgi, Camillo, 57, 58
Gordon, Deborah, 40, 41
Grassberger, Peter, 16, 111
Gravity, 20, 23, 25, 29, 30

Haldane, Andrew, 52, 53
Halley's comet, 20
Hierarchical predictive coding, 60, 61
Hierarchy, 81
    in life sciences, 81
    of complexity, 33, 124
    of information processing, 60, 61
    of neural processing, 60
    of organisation/structure, 28, 56, 58, 81, 84, 118
History, 10, 11, 23, 29, 30, 32, 33, 66, 73, 81, 82, 88, 110, 111, 115, 116, 123, 125, 128, 130
    evolutionary, 42, 57, 58, 82
    of complexity science, 11, 12, 21, 63, 73
Holland, John, 17, 82
Holldobler, Bert, 37
Honeybees, 41, 42, 57, 66, 96
Huberman, Bernardo, 17

Ideal gas, 22, see also Law.
Idealisation, 5
Immune system, 1, 17, 83, 85, 118, 124
Information processing, 6, 9, 40, 42, 44, 58–61, 65, 75–77, 88, 119, 126
Information theory, 8, 76, 77, 82, 89, 90, 97, 130, 135
Inhomogeneity, 24, 69, 89, 127
Invisible hand, 45

Jeong, Hawoong, 106

Kahnemann, Daniel, 48
Kauffman, Stuart, 16, 17, 125
Kertesz, Janos, 49
Kleiber, Max, 99
Kondor, Imre, 49

Langton, Christopher, 16, 17
Law, 3, 6, 7, 13, 14, 20–23, 67, 75, 77, 79, 103, 118, 121, 126, 128
    Boyle's law, 3
    emergence of, 74, 75, 80
    Hubble's law, 30
    ideal gas laws, 3, 5, 22, 28, 74, 84, 127
    Kepler's laws, 22, 75
    Newton's laws, 13, 14, 20, 22
    of co-existence, 22
    of gravitation, 5, 20
    of large numbers, 43
    of supply and demand, 45, 46
    of the pendulum, 5, 20
    Pareto's law, 46, 55
    power law, see Distribution
    second law of thermodynamics, 123
Life, 2, 8, 19, 25, 33, 76, 81–83, 116, 120, 121, 125, 127, 128, 130
Linearity, 50, 77, 78
Lloyd, Seth, 110, 111, 114, 115, 124
Log-log plot, 100
Logical depth, 82, 87, 140
Logistic map, 14, 100, 113
Lorenz, Edward, 14
Los Alamos National Laboratory, 16, 17
Lotka-Volterra model, 93, 102

MacKay, Robert, 78

Macrostate, 71
Mainzer, Klaus, 77
Mandelbrot, Benoît, 49, 108
Mantegna, Rosario, 49
Markets, 2, 6, 44, 45, 48–50, 61, 70, 71, 83, 124
Markov chain, 95, 136
Markov model, 112, 136, 137
Matrix, 108
    adjacency, 140
    Jacobian, 102
    stochastic, 95, 112
    transition, 112
Maxwell's equations, 22
Maxwell's theory of electromagnetism, 20
Maynard Smith, John, 17
Measures
    of complexity, 87, 88, 123
    of correlation, 96, 97, 138
    of disorder, 89, 91
    of diversity, 92
    of feedback, 93
    of history, 82, 110
    of memory, 110
    of modularity, 107
    of nested structure, 107
    of non-equilibrium, 95
    of nonlinearity, 99
    of numerosity, 88
    of order, 89, 91, 96
    of predictability, 138
    of randomness, 139
    of robustness, 101
Memory, 10, 66, 73, 81–83, 85, 110, 113, 125, 130
    in the brain, 59
    of an ant colony, 40
Microstate, 71
Milgram, Stanley, 106

Mitchell, Melanie, 84, 85
Modularity, 1, 10, 52, 53, 66, 68, 73, 81, 83, 107, 108, 126, 130
'More is different', 3, 9, 25, 60, 88, 128, 129
Morgenstern, Oskar, 47
Mutual information, 97, 98, 101, 138

Nested structure, 10, 66, 68, 73, 81, 83, 107–110, 124, 128, 130
Networks, 6, 9, 54–57, 73, 75, 76, 78, 83, 85, 88, 106, 107, 109, 119, 126, 137, 140
    average path length, 90, 91, 106
    clusters in, 107
    collaboration, 55
    financial, 53, 106
    genetic regulatory, 102
    IT, 71, 124
    metabolic, 55
    neural, 21, 106
    protein, 55, 106, 107
    scale-free, 55
    social, 5, 106, 107
    transportation, 124
Neurons, 6, 8, 44, 57–60, 66, 70, 79, 82, 83, 92, 98, 105
Newman, Mark, 108
Newton, Isaac, 5
Nihilism about complex systems, 117–119
Non-equilibrium, 10, 50, 65, 71, 85, 95, 96, 122, 130
Nonlinear dynamical systems, 100
Nonlinearity, 10, 14, 15, 46, 64, 65, 73, 77–79, 83, 85, 93, 99–101, 105, 108, 109, 128–130
Numerosity, 10, 61, 65–68, 73, 85, 88, 126, 130

Orbits of Mercury, 20

Order, 4, 8–10, 22–29, 31, 45, 57, 65, 68–74, 76, 77, 79–82, 84, 85, 88, 93, 96–98, 101, 111, 113–116, 119, 121–123, 125–128, 130

Packard, Norman, 17
Page, Scott, 93
Pagels, Heinz, 111
Particles, 3, 5, 19–22, 26, 29, 30, 32, 66–68, 74, 75, 79, 118, 120
Perelsen, Alan, 17
Phase transitions, 3, 16, 19, 24–29, 50, 67, 68, 72, 75, 101, 103, 104, 107, 127, 128
Physics, 2–4, 11, 12, 16, 17, 19–25, 29, 31, 33, 49, 50, 61, 70, 74, 75, 84, 100, 103, 118, 121, 122, 127, 128
    astrophysics, 3, 8, 33, 120
    condensed matter physics, 22, 25, 28, 66, 71, 72, 76, 79, 85, 97, 123, 126, 127
    statistical physics, 8, 21, 28, 125
Pines, David, 16
Poincaré, Henri, 14
Population growth, 14
Power law, *see* Distribution.
Pragmatism about complexity science, 119
Prediction, 7, 10, 11, 13, 14, 20, 35, 37, 60, 61, 66, 98
Probability distribution, 89–92, 95, 99, 100, 105, 108, 112, 113, 135, 137
    joint, 91, 98, 101, 108, 135, 138
Probability theory, 73, 89, 126, 135
Puck, Ted, 16
Purpose, *see* Function (purpose)

Quantum

chemistry, 3, 23, 116
    classical limit, 67
    computation, 124
    dynamics, 78
    fields, 20, 22, 127
    gravity, 21
    mechanics, 22
    multiverses, 124
    numbers, 22
    theory, 21
Quorum decision making, 41, 43, 44, 61

Radiation, 2, 5, 19–21, 28–31, 34, 36, 46, 50, 72, 104
Ramón y Cajal, Santiago, 57
Ramsey, Norman, 16
Randomness, 59, 89, 114, 124, 137, 139, 140
Reaction-diffusion model, 73
Realism about complex systems, 120, 125
Reductionism, 77, 84
Relativity, 21
Resilience, 79, 101, 103, *see also* Robustness.
Rich-get-richer effect, 46, 109
Robustness, 1, 10, 56, 66, 70, 73, 79, 80, 83, 85, 101, 106, 107, 110, 124–130, *see also* Resilience.
    functional, 59, 83, 101
    of algorithm, 101
    structural, 101, 107
Rosenblueth, Arturo, 12, 13
Rota, Gian-Carlo, 16

Sandpile model, 27
Santa Fe Institute, 17, 73
Scale
    characteristic scale, 68

length scale, 24, 28, 29, 31, 51, 68, 75, 77, 104, 114, 128
   time scale, 2, 3, 10, 21, 24, 28, 30–33, 35–37, 46, 49, 59, 65, 68, 70, 79, 80, 82, 93, 94, 103, 104, 127, 128
Scale invariance, 24, 26, 105
Schuster, Peter, 17
Scientific Revolution, 11
Scott, Alwyn, 16
Self-organisation, 4, 12, 16, 19, 27, 29, 37, 76–79, 84, 96, 113, 123, 127, 129, 130
Self-organised criticality, 27, 50, 105, 106, 123, 128
Self-similarity, 123
Sensitive dependence on initial conditions, 14
Shannon entropy, *see* Entropy.
Shannon, Claude, 137
Simon, Herbert, 81, 115, 124
Singer, Jerome, 16
Six degrees of separation, 106
Small-world effect, 55
Smith, Adam, 45
Smith, Vernon, 48
Snowflakes, 23
Stability analysis, 102, 103
Stanley, Eugene, 49
Statistical complexity, 87, 110
Statistical mechanics, 22, 24, 26, 72, 76, 97, 103, 111, 114, 121, 126
Sugarscape model, 94
Superposition principle, 14
Symmetry, 24, 69, 77, 81
Symmetry breaking, 24, 32, 77, 127

Thermodynamic depth, 124
Thermodynamics, 24, 25, 72, 95, 122, 123, *see also* Law.

Tipping points, 103, 106
Toffoli, Tommaso, 16
True measure complexity, 111, 115
Truisms, *see* Complexity.
Turing machine, 119
Turing pattern, 73
Turing, Alan, 73, 119, 139, 140
Tversky, Amos, 48

Ulam, Stanislaw, 15
Universal behaviour, 6, 9, 29, 75, 85, 88, 119, 126
Universe, 1, 19, 23, 28–33, 76, 82, 85, 116, 121, 123
Utility function, 47

Variance, 89–93, 97, 100, 136
von Bertalanffy, Ludwig, 13
von Neumann, John, 15, 47

Waggle dance, 42, 43
Water, 2, 7, 21, 23–26, 28, 29, 33–35, 42, 43, 68, 74, 80
Weather, 2, 14, 32–34, 36, 68, 127, 129
Wiener, Norbert, 12, 13
Wilczek, Frank, 16
Wilson, Edward, 37
Wolfram, Stephen, 16
World Wide Web, 46, 53–56, 90, 106, 109

Young, Karl, 111

Zhang, Yi-Cheng, 49